International Commercial Satellite Communications

Marcellus S. Snow

The Praeger Special Studies program—utilizing the most modern and efficient book production techniques and a selective worldwide distribution network—makes available to the academic, government, and business communities significant, timely research in U.S. and international economic, social, and political development.

International Commercial Satellite Communications

Economic and Political Issues of the First Decade of INTELSAT

PRAEGER SPECIAL STUDIES IN INTERNATIONAL ECONOMICS AND DEVELOPMENT

Praeger Publishers New York Washington London

Library of Congress Cataloging in Publication Data

Snow, Marcellus S
International commercial satellite communications.

(Praeger special studies in international economics and development)
Bibliography: p.
1. International Telecommunication Satellite Organization. 2. Artificial satellites in telecommunication--Economic aspects. I. Title.
HE9719.S57 384.54'56'0621 75-8410
ISBN 0-275-01150-X

PRAEGER PUBLISHERS
111 Fourth Avenue, New York, N.Y. 10003, U.S.A.

Published in the United States of America in 1976
by Praeger Publishers, Inc.

Printed in the United States of America

For Jo

PREFACE

The idea for this study arose during my two-year period of employment as a financial and political analyst with the Communications Satellite Corporation in Washington. Indeed, it was the desire to have the time and analytical wherewithal to undertake such a study that contributed in some measure to my subsequent decision to pursue a Ph. D. in economics at the University of California, Berkeley, and a career in academic teaching and research.

This book is based on my 1974 doctoral dissertation, "Cost Structure and Charging Policy in International Commercial Satellite Communications." It differs from it, however, in that I have thoroughly updated all statistical work to account for the one-year gap and have made other changes as well.

In this study I have tried to strike an agreeable compromise between too narrow and too broad focus. Although the study is written by an economist and is based largely on the opus which represents an acolyte's rite of passage into the economics profession, I hope it will appeal to noneconomists as well. My first professional contact with INTELSAT and satellite communications, in fact, predated my graduate studies in economics but followed closely my completion of a master's degree in international relations at the Johns Hopkins University School of Advanced International Studies. I hope that this broader interest is evident in the structure of the study, and that discussions and conclusions of a noneconomic nature will be taken at least as seriously as my t-statistics.

The fact remains that this is a book which deals with fundamentally economic issues in an area which has generated more commentary from other disciplines, such as law and political science. It is, I am sure, the first book-length study of INTELSAT by an economist and the first to make extensive use of INTELSAT financial data and other official documents. I hope that it will not be the last, and this work will have met at least some measure of success if it stimulates contributions from those wishing to correct my errors or to flesh out areas I have left relatively untouched. There is, alas, ample opportunity to do both.

Part I is a simple description of the origins of INTELSAT, its organization and technology, and the basic issues relating to its first decade of operation. Part II, the most unremittingly economic treatment in the study, derives cost functions for INTELSAT on the basis of ten years' operating data,. and documents and analyzes in considerable detail the various criteria of charging policy which INTELSAT has followed, whether explicitly or by default. The source of data may be novel, but the issues involved are the familiar ones of regulatory policy in the presence of scale economies, excess supply, Averch-Johnson effects, and so on. Part III deals with ostensibly economic issues such as natural monopoly, scarce resource constraints, and economic development, but in a way that presumes no rigorous knowledge of the considerable economic literature bearing on those areas. It is more an attempt to place INTELSAT in a meaningful context within that literature than to deal with any one topic conclusively. Part IV attempts a summing up, once more from an economist's standpoint. I argue that INTELSAT has been doing a tolerably good job based on the criterion by which it should properly be judged, which is the interest of the consumers in its various member countries.

This study draws on comments and generous help from many individuals, but those I am about to thank are in no way responsible for its shortcomings. I am particularly grateful to Theodore E. Keeler, my dissertation chairman at Berkeley, for the care, thoroughness, and dispatch with which he helped me complete my dissertation. For suggestions contributing to the improvement of this study and its predecessor I am indebted to Joe Bain, Albert Fishlow, Darius Gaskins, Bent Hansen, Daniel McFadden, C. Bartlett McGuire, Stephen Peck, Joe Pelton, and Bill Snow. More generally I wish to thank my former Comsat associates in Washington and Geneva, as well as my colleagues in the Department of Economics at the University of Hawaii, for providing environments in which my ideas could take shape and be refined.

I appreciate the financial aid extended by a Dissertation Research Grant awarded by the Institute of Public Utilities, Michigan State University.

Finally, for quite different reasons, I am grateful to my wife, to whom this study is dedicated, and to our two children.

CONTENTS

Page

LIST OF TABLES AND FIGURES

PART I

INTRODUCTION

CHAPTER 1
SATELLITE COMMUNICATIONS AND INTELSAT

THE SETTING OF THE PROBLEM

The aim of this study is to evaluate the performance of international commercial satellite communications with respect to economic efficiency. The source of information is limited: INTELSAT (the International Telecommunications Satellite Organization), established in 1964 and now comprising nearly 90 member states, has been and remains the only source of worldwide commercial satellite service. For that reason, data will be drawn primarily from INTELSAT's operating records, and conclusions will of necessity be based on those data. The present study, therefore, while resembling in some aspects examinations of more conventional public utilities such as electricity generation, will lack a cross-sectional analysis of dozens of firms and a time series of several decades of operating data. Many of the same kinds of questions, such as those regarding cost and price structure, will be raised and examined in this study, but the relative scarcity of data will limit not only the power of the statistical methods used, but also the certainty of verbalized arguments.

Despite these limits on data and conclusiveness, an inquiry into the economic aspects of international satellite communications is viewed as being of interest and utility. In the first place, there is no existing overall study of the economic aspects of international satellite communications in which primary data are utilized, as in the present study. Second, satellite communications are novel in a number of ways that invite investigation and promise to offer new perspectives on traditional questions of public utility and welfare economics. Third, the international institutional framework which was established to accommodate this new technology is also unprecedented, and has in fact been studied at some length.

The existence of studies of INTELSAT in allied fields will facilitate the present inquiry into the economic aspects of satellite communications. INTELSAT is unique as an international organization with a specifically commercial mission that must meet a large operating budget out of revenues derived from its services. Moreover, new questions of law, particularly regarding the international use of outer space, the regulation of the radio frequency spectrum and of orbital parking spaces for satellites, control of direct satellite broadcasting to foreign countries, and international patent and data agreements, have arisen directly out of INTELSAT's brief experience. INTELSAT's present organization and charter have emerged from two grueling negotiating sessions; the first session established two interim agreements for INTELSAT in 1964[1] and the second produced two permanent agreements signed in 1971[2] after two years of negotiations. While adequate and careful studies of INTELSAT exist in a number of related areas, as noted above, no such examination exists in regard to the economic aspects of international satellite communications, and it is this gap that the present study aims to fill.

The question of narrowness of focus is an important one and will affect the types of conclusions reached in this study. To arrive at welfare conclusions, one would want as broad a perspective as possible. For example, satellites are only one mode of telecommunications, undersea cables being a close substitute and mail delivery a more distant one. Even transportation competes slightly: perhaps private-line television by satellite could substitute for international travel. Yet a study encompassing all these facets of international communication would be intractably broad.

The decision to focus this study on INTELSAT will entail less than adequate attention to many other aspects of satellite communications. For example, the most telling omission may be that of the earth stations, which are not part of INTELSAT but are instead generally owned and operated by the national administrations of the countries in which they are located. The terrestrial distribution networks, which are necessary to bring satellite-relayed channels to the final consumers at each end, are also omitted. In addition, the role of cables in telecommunication is not examined in detail, although it is addressed briefly in Chapter 8.

These omissions prevent, to a certain extent, the possibility of arriving at conclusions as broad as might be felt desirable, since there are technological and economic interrelationships between the space segment of INTELSAT (the focus of the present study) and areas outside the scope of this study. For example, the weight (and therefore the capacity) of satellites has been constrained by the thrust of available launching vehicles provided for INTELSAT by the U.S. National Aeronautics and Space Administration (NASA). In addition, there is an inverse relationship between the size and expense of satellites and the size and expense of the earth stations with which they communicate.

Furthermore, the presence or absence of cables along particular routes will obviously affect the demand patterns for satellite communications.

While these areas impinging on the INTELSAT space segment will be mentioned, discussed, and to some extent examined in the course of this study, they will not form the basic core of analysis. If they did, the study would either expand beyond feasible limits or increase the breadth of coverage primarily at the expense of depth. For example, one might attempt to include earth stations in the analysis, since they are an integral part of establishing a satellite circuit. But each country has its own station or stations, each is in a different institutional and financial environment, and each has its own cost curves and is equipped differently. Yet the INTELSAT space segment, which is used by all the earth stations, is a single entity, administratively and financially, and the gain in scope to be obtained by including detailed cost data from each of the more than sixty existing earth stations would be more than outweighed by the added complexity of analysis that would be required. Such analysis would easily have exceeded the scope of time and resources that were available for the present study. The same is even more applicable for still broader levels of analysis, such as telecommunications in general. To the extent possible, of course, conclusions reached based on the INTELSAT space segment data will be qualified by comments relating to broader questions regarding areas such as earth stations, competing modes such as cables, and so on.

In focusing analytic detail on the INTELSAT space segment, the following main questions will be posed and discussed. Part II will first derive several sets of total, marginal, and average cost curves for INTELSAT operation; these will then be used as a basis for determining optimality properties of INTELSAT's pricing policy as it evolved and as it is presently constituted. In this discussion, the welfare criterion will be that of the literature on marginal cost pricing, its disadvantages, and optimum deviations therefrom as second-best solutions. Price discrimination and regional pricing will be singled out as possible welfare-increasing alternatives to the present average-cost, "revenue requirements" approach to recovering operating expenses, which INTELSAT has borrowed from conventional public utility practice. Decreasing average costs (increasing returns), for which evidence is presented in Part II, will be found to have a strong influence on the types of pricing policies found optimal or quasi-optimal as second-best solutions, due to the inability of marginal cost pricing to recover operating expenses (average costs) in the presence of decreasing average costs. Part III will take the broader perspective spoken of above by discussing in less detail a number of additional economic issues raised by satellite communications. Finally, Part IV will discuss welfare criteria implicit in INTELSAT's performance and will suggest various alternatives while rejecting others.

THE FRAMEWORK: HISTORY AND STRUCTURE

This section will present a highly selective sketch of the technological requirements of satellite communications, as well as some historical and institutional details relating to the establishment and development of INTELSAT as the world's primary supplier of international satellite communications. Particular attention is called to footnoted references as well as the bibliography at the end of this study for greater detail. Journalistic, legal, and structural treatments of INTELSAT and satellite communications in general are much more abundant than economic treatments, and no effort will be made here to duplicate the former treatments.

Technology

Artificial earth satellites were envisaged at least as early as 1945, and Russia's first Sputnik, launched in 1957, was the initial realization of earlier theorizing. The United States countered soon by launching the first true communications satellite in 1958.

A primary constraint on the size and capacity of satellites has been the power of launchers, as noted above. For this reason, the early satellites were very light, with highly elliptical orbits, approaching within a few hundred miles of the earth at their perigee. This meant they had to be tracked by earth stations each time they passed across their line of sight. Early plans for the INTELSAT system presupposed satellites with elliptical or "random" orbits, and the Soviet Union's domestic system is to this day based on random-orbiting satellites, which do furnish some relative advantages for northerly latitudes.

At an altitude of 22,300 miles above the equator, the force of gravity is such that a body will rotate with the same angular velocity as the earth, appearing to remain stationary above a particular point on the equator. A satellite at that altitude is visible to earth stations on about one-third of the earth's surface, excluding the north and south polar regions. Powerful rocketry evolved over time to launch all INTELSAT satellites into these so-called synchronous or geostationary orbits, rather than elliptical orbits. INTELSAT satellites are equippped with apogee motors that are fired when the satellite reaches the apogee of the elliptical orbit into which it is first launched by a NASA rocket. From the apogee, it goes into synchronous orbit. From that time on, the satellites are maneuverable by means of jets of fuel that can be fired to give an impulse by which the satellites can be drifted from position to position around the geostationary orbit. This fuel supply is limited, and there is some risk involved in the drifting maneuver.

Furthermore, the satellites cannot presently be reached in orbit for repair and are consequently vulnerable to breakdowns that could be easily repaired on the earth.

A given combination of power and bandwidth is required to establish communications channels between earth stations. The basic unit of measurement of communications capacity used by INTELSAT for technical and financial purposes is the so-called unit of utilization,[3] which measures the amount of power and bandwidth needed to establish one-half of a voice (telephone) conversation between two "standard" earth stations in the INTELSAT system, that is, stations furnishing a given amount of power to the satellite. The unit of utilization corresponds to a "voice channel," or one-half of a voice circuit, which is a two-way telephone conversation. Allotments of satellite capacity are generally arranged bilaterally between corresponding administrations and then applied for from INTELSAT. Satellite capacity can be used for purposes other than voice, such as data and television. Television is perhaps the most dramatic use of satellites, because a television channel occupies several hundred voice channels and cannot be carried on cables due to their limited capacities. Therefore, the first live trans-oceanic television came in the early 1960s by experimental satellites developed by the American Telephone and Telegraph Company (AT&T) and the U.S. government.

Satellites differ from cables in many ways. Primarily, satellites require no physical connection between ground terminals, and utilize the radio frequency spectrum, in contrast to cables. Satellite costs are independent of the distance between corresponding ground stations, which is obviously not the case for cables. Such major differences between these two modes will affect conclusions in this study regarding the price and cost structure of satellite communications.

INTELSAT has had four different series of satellites. The first, called INTELSAT I and also Early Bird, consisted of only a single launched satellite, which was put into orbit in 1965. It provided 480 units of utilization across the Atlantic and could connect only two earth stations at a time. INTELSAT II satellites were first launched in 1966 and provided the same amount of capacity but offered multidestination reception by receiving earth stations. A major purpose of this series was to provide capacity to NASA for the Apollo lunar exploration program. The INTELSAT III series, with an initial launch in 1968, provided the first truly global coverage, with satellites over the Atlantic, Pacific, and Indian Ocean regions. In contrast to all other INTELSAT satellites, which were built by Hughes Aircraft Company as the prime contractor, the INTELSAT III series was built by TRW Systems, another aerospace contractor which, like Hughes, had had considerable prior experience building military and experimental satellites for the U.S. government. Each INTELSAT III satellite had 2,400 units of capacity. INTELSAT IV satellites, first launched in 1971, have a maximum capacity of 10,000 units each and the option of spot-beam usage, by means of which narrow beams of high power are focused onto small areas of the earth's surface, as well as the usual global beams which cover larger areas at lower power levels. INTELSAT is preparing for usage a successor series of INTELSAT IV-A satellites, modified by Hughes from the earlier IV series, as an alternative to a more advanced INTELSAT V series.[4] The first of six IV-A satellites is scheduled for launch in late 1975; in July 1975, the INTELSAT Board of Governors issued a request for proposals for the INTELSAT V satellites indicating early 1979 as the expected launch date for that series.

History

Trials with Echo, Telstar, and other experimental communications satellites in the early 1960s[5] indicated their commercial promise and led to discussions in the U.S. Congress on the institutional and financial framework in which the new technology should be allowed to bear fruit. The legislative events paving the way for the Communications Satellite Act of 1962[6] involved, in essence, alternative proposals for three main types of commercial structures: a government-owned corporation, an all-private corporation dominated by the U.S. overseas carriers, and a hybrid corporation with ownership to be shared by private investors, the U.S. overseas communications carriers, and the U.S. government. The shared ownership concept won out, more as a matter of compromise, in the eyes of many observers, than for its intrinsic merits. In the 1962 act, the Communications Satellite Corporation (Comsat) was chartered by Congress; ownership was divided between the public at large and the major U.S. overseas carriers; appointment of some members of the board of directors by the U.S. president was mandated; and controls and regulatory powers to be exercised by the Federal Communications Commission (FCC) and the Department of State were established.

Although some European countries had already built earth stations to participate in earlier transatlantic satellite broadcasts of an experimental nature, Europe was well behind the United States in satellite and launcher technology. Comsat, acting from its position of strength, negotiated with European governments to set up an international satellite system but appeared to prefer bilateral arrangements along former cable patterns to the establishment of a new international organization. The European countries formed a common bloc by organizing the CETS (French initials for European Telecommunications Satellite Committee) to increase their negotiating power. The final result was the two abovementioned interim agreements—the INTELSAT Interim Agreement and Interim Special Agreement—signed in August 1964 by a number of countries, to which several dozen more later acceded. The agreements established INTELSAT (the name INTELSAT was actually not coined until somewhat later) as an international organization governed by the Interim Communications Satellite Committee (ICSC); the ICSC had about 18 members, each representing countries or groups of countries with 1.5 percent or more of the projected 1968 INTELSAT traffic level (Article IV(b) of the Interim Agreement). Advisory subcommittees with powers of recommendation were provided for (Article IV(c)-(d)); these came to include the Advisory Subcommittees for Technical, Financial, and Contracting Matters (ICSC/T, ICSC/F, and ICSC/C, respectively). INTELSAT headquarters were established in Washington. Most importantly, the interim agreements named Comsat as the "Manager" of INTELSAT (Article VIII), meaning that Comsat personnel, aside from pursuing the corporate interests of Comsat, were from time to time put on loan to INTELSAT, which itself had no full-time employees as such. Comsat was paid by INTELSAT, on a cost-reimbursable basis, for those services and received an annual management fee as well. Some INTELSAT members have charged that this arrangement put Comsat in a conflict-of-interest situation in which Comsat's interests could be furthered by its employees ostensibly working on behalf of INTELSAT.

Because of their lack of bargaining power and Comsat's dominant role in the new organization, other INTELSAT members were successful in insisting that provision be made in the 1964 agreements for "permanent" or "definitive" arrangements after a number of years,[7] by which time it was assumed that a renegotiation of the agreements would diminish U.S. influence. Thus the 1964 agreements were known as "interim" agreements. By late 1968, as the result of ICSC meetings devoted exclusively to bringing about permanent agreements, recommendations were in hand for renegotiation of the INTELSAT agreements. In 1969, the largest international conference ever held in Washington began; negotiations were expected to last a few weeks or months at most, but they lasted two years, including pauses during which working groups carried out specialized tasks. By May 1971, the proposed permanent agreements were opened for signature. They entered into force in February 1973, forming the basis of INTELSAT's present structure and operation.

Comsat was in a unique position domestically and internationally.[8] As noted, the U.S. international carriers were its principal stockholders, and the presence in particular of AT&T executives on the Comsat board of directors put AT&T in a conflict-of-interest situation because of its competing cable ownership. The market structure imposed on telecommunications by the Communications Satellite Act, it has been argued,[9] adversely affected domestic performance by distorting the incentives for satellite development and use. In addition, carrier pressure on the FCC succeeded in foreclosing the retail market to Comsat, forcing it instead to wholesale satellite circuits to the international carriers as a "carrier's carrier." Finally, Comsat was overcapitalized in its early years because the costs of the synchronous INTELSAT system were lower than the random-orbiting costs which had been anticipated, so it looked to a domestic U.S. system as an outlet for its surplus capital. But Comsat was denied a substantial share of the domestic U.S. satellite system, first by a 1970 White House statement favoring limited competition in that area,[10] and then by later FCC decisions awarding domestic system approval to non-Comsat entrants. (For a discussion of the U.S. domestic system see Chapter 6.) At present, however, Comsat no longer has carrier stockholders, and does have approval for participation in domestic satellite communications. Finally, Comsat had originally wanted to own all U.S. ground stations but was again thwarted by pressure from the carriers, which succeeded in inducing the FCC to order that the earth stations be owned jointly by Comsat and the U.S. carriers with Comsat as manager.[11]

Present Structure

In its present structure, as established in the permanent agreements (entering into force early in 1973), INTELSAT is a cooperative in the sense that its users and owners are identical. Some exceptions, such as use of the system by nonmembers, are provided for. Furthermore, under the definitive arrangements, ownership quotas are periodically realigned with recent usage percentages.[12] Since INTELSAT is required to cover its costs and pay a rate of return of 14 percent to those providing its capital, its members can have conflicting interests. As users, they want usage fees to be as low as possible. As owners, however, they want the maximum return on their investments. As will be shown in the Appendix to this study, if usage is exactly equal to ownership, the rate of return is irrelevant in the sense that higher usage fees to achieve a higher return would exactly counterbalance the higher return obtained. If usage exceeds ownership, an INTELSAT member will have an incentive for as low a return as possible, since he is a net user. If usage is less than ownership, he is a net owner and will want as high a return as possible. Since the interim agreements did not provide for a periodic realignment of ownership quotas to reflect recent past usage, but merely established initial ownership quotas based on projected 1968 usage, serious imbalances of ownership and usage arose for a number of member countries in the later years of the interim regime and often caused serious conflicts in the ICSC concerning the rate of return, the unit of utilization charge, depreciation policy, and other financial matters.

According to the INTELSAT Agreements, the present INTELSAT organization is broader and more representative of views of smaller countries than it was during the interim period. INTELSAT still has an executive body (Articles IX and X), now called the Board of Governors rather than the ICSC, which votes according to ownership quotas, although no country is allowed more than 40 percent of the vote. In addition, there are two wholly new bodies: the Assembly of Parties (Article VII), or sovereign states, and the Meeting of Signatories (Article VIII), or operating agencies. Each of these has limited though substantive powers to influence INTELSAT policy; decisions in these two bodies are taken on a one-man, one-vote basis and thus increase the influence of smaller users, which are primarily less developed countries. Finally, INTELSAT was assigned an "executive organ" of its own permanent employees which is not entirely specified in the new agreements. During the six-year period from 1973 to 1979, INTELSAT will have a secretary general with a small full-time staff (Article XII), and Comsat will function as a "management services contractor" with specified duties (Article XII(e)). After the six years, there will be a director general with a staff structure to be decided upon (Article XI). Beginning in 1979, INTELSAT may choose management contractors other than Comsat, although Comsat is not foreclosed from continuing its relationship thereafter (Article XI(c)).

Operation of INTELSAT

Details of INTELSAT's operation are specified in the separate Operating Agreement. Financial transactions between signatories (that is, sale or purchase of investment shares) and between signatories and INTELSAT (initial membership quotas, withdrawal, etc.) are given in some detail in Articles 7 and 21. Operational matters such as earth station standards (Article 14) and allotments of satellite capacity (Article 15) are dealt with. Legal matters such as procurement (Article 16), liability (Article 18), and acquisition of inventions and technical information (Article 17) by INTELSAT are regulated. Principles of financing and charging policy are laid down in Article 8. In additions, the mechanics of determining investment shares based on recent past usage are spelled out (Article 6).

NOTES

1. U.S. State Department, Treaties and Other International Acts Series 5646, _Agreement Establishing Interim Arrangements for a Global Commerical Satellite System, and Special Agreement,_ Washington, August 20, 1964 (hereafter cited as _INTELSAT Interim Agreement_ and

Interim Special Agreement). One of those agreements was a general agreement, called the Interim Agreement, among governments of states, with those states signing the treaty technically referred to as parties to that agreement. The other was called the Interim Special Agreement, covering operating details of INTELSAT; it was signed by the governments or by operating agencies designated by the governments, called signatories. Articles of the Interim Agreement given in Roman numerals, and those of the Special Agreement in Arabic numerals.

2. U.S. State Department, Treaties and Other International Acts Series 7532, International Telecommunications Satellite Organization (INTELSAT), Agreement Between the United States of America and Other Governments, and Operating Agreement, with Annex Concluded by Certain Governments and Entities Designated by Governments, Washington, August 20, 1971 (hereafter cited as INTELSAT Agreement and INTELSAT Operating Agreement, respectively, with the Roman/Arabic distinction the same as in the interim agreements).

3. For the current technical definition of a unit of satellite utilization accepted by INTELSAT, see INTELSAT Tariff Manual, April 3, 1974, pp. 3-4. This definition depends in turn on definitions adopted by the International Telecommunication Union (ITU) and its subsidiary bodies.

4. See Communications Satellite Corporation, Pocket Guide to the Global Satellite System, March 1974, for a listing of INTELSAT earth stations, a brief description of each generation of INTELSAT satellites, a roster of INTELSAT members, and a record of all INTELSAT launches.

5. See, for example, Joseph N. Pelton, Global Communications Satellite Policy: INTELSAT, Politics and Functionalism (Mt. Airy, Md.: Lomond Books, 1974), pp. 41-63.

6. U.S. Congress, An Act to Provide for the Establishment, Ownership, and Regulation of a Commercial Communications Satellite System, and for Other Purposes, Public Law 624, 87th Cong., 2d sess., August 31, 1962.

7. INTELSAT Interim Agreement, Art. IX. This article provided that no later than January 1, 1969, the ICSC was to submit a report concerning "definitive arrangements for an international global system which shall supersede the interim arrangements established by this Agreement," with January 1, 1970, set as the date for their entry into force. In fact, the date of entry into force was February 12, 1973.

8. For structural treatments of the U.S. domestic telecommunications industry, see Kurt Borchardt, Structure and Performance of the U.S. Communications Industry (Boston: Harvard Business School, 1970); and Manley R. Irwin, The Telecommunications Industry: Integration vs. Competition (New York: Praeger, 1971).

9. See, for example, Kenneth Bishop Stanley, "International Telecommunications Industry: Interdependence of Market Structure and Performance Under Regulation," Land Economics 69, no. 4 (November 1973), pp. 391-403.

10. U.S., Office of the President, "Memorandum for the Honorable Dean Burch, Chairman of the Federal Communications Commission," January 23, 1970, signed by Peter Flanigan, assistant to the president.

11. Legal and journalistic treatments of these early questions regarding Comsat's evolution in domestic regulation and legislation are numerous but uneven in quality. Three articles in _Fortune_ are of interest: John McDonald, "The Comsat Compromise Starts a Revolution" _Fortune_, October 1965; Charles E. Silberman, "The Little Bird that Casts a Big Shadow," _Fortune_, February 1967; Lawrence Lessing, "Cinderella in the Sky," _Fortune_, October 1967; Harvey J. Levin, "Organization and Control of Communications Satellites," _University of Pennsylvania Law Review_ 113, no. 3 (January 1965), pp. 315-57; Herman Schwartz, "Govermentally Appointed Directors in a Private Corporation—The Communications Satellite Act of 1962," _Harvard Law Review_ 79, no. 2 (December 1965), pp. 350-64; Herman Schwartz, "Comsat, the Carriers, and the Earth Stations: Some Problems with 'Melding Variegated Interests,'" _Yale Law Journal_ 76, no. 3 (January 1967), pp. 441-84; Richard R. Colino, "INTELSAT: Doing Business in Outer Space," _Columbia Journal of Transnational Law_ 6, no. 17 (1967), pp. 17-60; Blander treatments include George E. Ashley, "International Communications: What Shape to Come?," _Law and Contemporary Problems_ 34, no. 2 (Spring 1969), pp. 417-28; Asher H. Ende, "International Telecommunications: Dynamics of Regulation of a Rapidly Expanding Service," _Law and Contemporary Problems_ 34, no. 2 (Spring 1969), pp. 390-416; and O. W. Riegel, "Communications by Satellites: The Political Barriers," _Quarterly Review of Economics and Business_ 2, no. 4 (Winter 1971), pp. 23-35.

12. _INTELSAT Operating Agreement_, Art. 6.

PART II

INTELSAT'S COST AND TARIFF STRUCTURE

CHAPTER

2

DERIVATION OF TOTAL, AVERAGE, AND MARGINAL COSTS

In the short run, with which empirical cost derivation must necessarily first concern itself, fixed costs are those which remain constant over the range of output, while variable costs vary with output.

In terms of the satellites of the INTELSAT system, the short run will first be conceptualized in terms of the productive capacity and lifetime of individual satellites of the different series, or generations, of INTELSAT satellites: INTELSAT I-II, III, and IV. Later in this chapter the short run will be extended to encompass particular configurations of satellites from each of those series as they have been employed by INTELSAT in the past over all three ocean regions.

VARIABLE COSTS

Short-run variable costs will be estimated statistically after a necessary conceptual dichotomization has been made along the lines of accounting costs. Since it began operation in 1965, INTELSAT has divided its costs into three main areas: capital costs (depreciation), return to capital, and current operating expenses. Capital outlays and the return paid to those providing that capital have of course been used to pay for the satellites, development costs, and launcher services used by INTELSAT. It seems clear that the capital cost of providing a launched satellite, together with return to capital used for that purpose, represent fixed costs in the short run, just as a factory, once built, is a fixed cost in the short run. Therefore, a rough assignment of "capital" costs to short-run fixed costs will be made. Three different accounting measures of "all other" costs, consisting mainly of current cash operating expenses, will be tested statistically to determine which is most sensitive to variations in output (but still insensitive to variations in capacity) and thereby the best proxy for variable costs.

TABLE 2.1

Three Possible Measures of INTELSAT's Variable Costs, and Traffic and Capacity Averages for 1965-74, Inclusive

Year	Departmental Expenses Before Capitilization ($000, 1965 Prices[a])	Departmental Expenses After Capitilization ($000, 1965 Prices[a])	Cash Operating Expenses ($000, 1965 Prices[a])	Total Traffic Units[b]	Average In-Orbit Capacity Units[d]
1965	4,684	4,621	4,889	141[c]	360
1966	3,781	3,726	3,923	160	480
1967	6,085	4,273	4,801	805	1,360
1968	7,968	5,751	6,659	1,420	1,920
1969	9,953	7,449	9,743	2,197	7,520
1970	10,003	7,368	10,358	3,698	11,680
1971	9,668	7,735	10,921	5,117	19,500
1972	9,511	8,269	10,952	6,713	41,167
1973	9,958	9,126	11,660	8,626	55,333
1974	10,785[e]	9,886[e]	12,628[f]	10,581	57,433

[a]Discounted by hourly wage rate for U.S. communications equipment industry.

[b]Twelve times monthly average of equivalent full-time unit-years.

[c]Annualized from six months' operation.

[d]Calculated from nominal capacities of in-orbit satellites still functioning and undepreciated.

[e]Imputed from estimated cash operating expenses figure for 1974.

[f]Based on budget figures for 1974.

Sources: Expenses before and after capitalization, and cash operating expenses, are drawn from INTELSAT Financial Reports (annual); traffic figures are from INTELSAT System Status Reports (monthly and bimonthly).

TABLE 2.2

Regression Results for Data in Table 2.1, Using Logarithmic Specification

Dependent Variable	Equation with Estimated Least-Square Parameters (t-values shown underneath[a])	R^2	F-Statistic[b]	D-W Statistic
DEBC	$e^{14.14}(TRAF)^{0.37}(CAP)^{-0.12}$ (72.21)** (3.14)* (-1.28)	.9206	40.58**	1.45
DEAC	$e^{14.20}(TRAF)^{0.02}(CAP)^{0.15}$ (76.39)** (0.21) (1.64)	.9155	37.92**	1.68
COE	$e^{13.96}(TRAF)^{0.08}(CAP)^{1.34}$ (60.50)** (0.58) (1.34)	.9209	40.75**	1.34

[a]For seven degrees of freedom, $t_{.05} = 2.365$ and $t_{.01} = 3.499$ for two-tailed test.
[b]For two and seven degrees of freedom, $F_{.05} = 4.74$ and $F_{.01} = 9.55$.
*Significant at the 95 percent level.
**Significant at the 99 percent level.

Variables: DEBC = Departmental expenses before capitalization
DEAL = Departmental expenses after capitalization
COE = Cash operating expenses
TRAF = Traffic
CAP = Capacity

Source: Compiled by the author.

Current operating expenses have consisted largely, but not exclusively, of so-called departmental expenses, that is, labor, overhead, and other expenses assigned directly to particular departments of INTELSAT's manager, Comsat. Some of these departmental expenses have been capitalized, but most have not. From these categories, there are three measures of costs in INTELSAT financial records that are the likeliest candidates for variable costs in the economic sense: departmental expenses before capitalization (DEBC); departmental expenses after capitalization (DEAC); and current operating expenses (COE). These categories, together with average equivalent annual full-time traffic (TRAF) and in-orbit capacity (CAP) figures, are presented in Table 2.1 for the ten-year period 1965-74.

The data in Table 2.1 were used to derive variable costs statistically A logarithmic and linear specification were examined separately for DEBC, DEAC, and COE as alternative dependent variables. Both traffic (output) and capacity (plant size) are first specified as explanatory variables. The linear form was specified without a constant term so that variable costs would be zero at zero capacity and traffic. It turned out, however, that the linear form was misspecified, as indicated by the nonrandom distribution of the actual values of the dependent variables about the estimated values. Further, one has an intuitive preference for a specification which will allow for increasing returns to scale. Therefore, only the logarithmic specification is presented in Table 2.2, with CAP and TRAF as the explanatory variables.

The best proxy for variable costs will be the one with a highly significant traffic coefficient (showing that it is sensitive to the level of output) and a capacity coefficient not significantly different from zero (showing that it is not sensitive to the level of plant size or capacity). Ideally, one would want to conduct cross-sectional regressions for a given plant size (in-orbit capacity) in order to better capture the theoretical essence of short-run variable costs, but data were not available for such an approach. As a next-best approximation, therefore, sensitivity to output and insensitivity to capacity are the criteria used to select the best accounting proxy for short-run variable costs.

It is clear from Table 2.2 that the first equation listed, with DEBC as its dependent variable, is the best of the three from this standpoint. It is the only one of the three in which TRAF appears with an estimated coefficient significantly different from zero, yet the capacity coefficient itself is still not significantly different from zero.

It should be noted that if there were no increasing returns to scale for traffic (output) level, that is, if the exponential coefficient of the traffic term in the DEBC regression were estimated greater than unity instead of 0.37, a U-shaped short-run average cost curve would be indicated.[1] The value of 0.37 indicates, by contrast, that the short-run average cost function will decline for all levels of feasible output.

Capacity was specified as an explanatory variable in Table 2.2 in order to help select the best accounting proxy for variable cost. Having selected DEBC for that purpose, however, it is now necessary to regress DEBC on traffic, or output, level only, since in theory this should depend solely on output. The logarithmic format is retained, and the regression results, based on columns 2 and 5 of Table 2.1, are

$$\underset{(70.74)}{\text{DEBC} = 1{,}439{,}752}(\text{TRAF})^{\underset{(8.59)}{0.22}}, \qquad (2.1)$$

where both coefficients are significant at the 99 percent level, and where the unit of measurement is 1965 dollars for costs and units of utilization for traffic. Equation 2.1 will serve as the short-run variable cost function in the development below. The removal of capacity from the explanatory specification, even though it was not found significantly different from zero, has reduced the exponential coefficient of traffic from 0.37 to 0.22. It should be emphasized that this coefficient cannot be interpreted merely as indicating very pronounced static economies of scale in variable costs. The regressions have been performed over a time series during which marked progress occurred in satellite technology, and much of the decrease in unit variable costs at higher output levels implied by equation 2.1 must surely be attributed to dynamic factors rather than to static scale economies in the conventional sense. This holds true even though variable rather than fixed (capital) costs are discussed here. The dynamic economies involved might derive from such sources as organizational and managerial innovation and learning-by-doing.[2]

FIXED COSTS

All costs not captured as variable in the treatment above will be considered as fixed in the short run. Fixed costs will therefore derive from three sources: capital expenditures for the different satellite series; payments of interest to providers of capital funds, assigned to the different satellite series; and current operating expenses less variable costs, that is, COE less DEBC, which captures residual unassigned costs, also assigned to the different series. Regarding capital expenditures, Table 2.3 presents a year-by-year and series-by-series breakdown in terms of 1965 prices.

After the unassigned tracking and control expenses are assigned to the different series in proportion to capital costs already disaggregated, a percentage of total capital expenditure figures is obtained for each series, as shown in Table 2.4. The total series figures are then broken down into successful launch figures and in-orbit satellite-per-

TABLE 2.3

Capital Expenditures for INTELSAT, by Series and Year, 1965 Prices
($000)

Year	INTELSAT I	INTELSAT II	INTELSAT III	INTELSAT IV	Tracking[b]	Control[b]	Year Total
1964	6,903	—	—	—	—	—	6,903
1965	9,468	—	—	—	—	—	9,468
1966	1,847	10,799	7,091	—	—	25	19,762
1967	—	17,267	17,292	—	209	17	34,785
1968	—	4,179	15,641	3,086	1,222	—	24,128
1969	—	1,184	23,867	26,749	3,952	—	55,752
1970	—	—	17,282	51,375	1,286	—	69,943
1971	—	—	29	59,521	1,131	—	60,681
1972	—	—	—	37,191	530	—	37,721
1973	—	—	—	21,604[a]	—	—	21,604
1974	—	—	—	3,713[a]	—	—	3,713
Totals	18,218	33,429	81,202	203,239	8,330	42	344,460

[a]INTELSAT projections through end of INTELSAT IV acquisition period (including tracking and control).

[b]Assigned by series for most years.

Note: Prices are deflated by Wholesale Price Index for electric equipment and machinery.

Sources: INTELSAT Financial Report (annual), Financial Status of INTELSAT (bimonthly) ICSC/F and BG/F Summary Records

TABLE 2.4

Percent of Capital Expenditures for Each Series

Series	Capital Expenditures ($000)	Percent of Total
INTELSAT I-II	52,936	15.4
INTELSAT III	83,220	24.1
INTELSAT IV	208,304	60.5
Totals	344,460	100.0

Source: Compiled by the author.

year figures, as shown in Table 2.5. In addition, the mid-range estimate for the planned INTELSAT V series is used to estimate corresponding capital cost figures for that series.

The next component of short-run fixed costs is payments to return on capital. Table 2.6 indicates discounted total payments in this category during the period 1965-74. The figure of 14 percent is chosen to deflate return to capital because it was that cumulative rate of return which investors in INTELSAT were told they would eventually attain—and eventually did attain. Therefore, the artificial assumption is made that capital for INTELSAT is raised in a special "capital market" for commerical satellite communications and that such capital costs 14 percent more each year than it did the previous year.

Dividing the amounts in Table 2.6 by the number of satellite-years per series given in Table 2.5, column 7: one obtains \$12,700,000 ÷ 12 = \$1,058,333 per satellite-year for INTELSAT I-II; \$19,875,000 ÷ 25 = \$795,000 per satellite-year for INTELSAT III; and \$49,893,000 ÷ 42 = \$1,187,929 per satellite-year for INTELSAT IV. To estimate total payments to capital per satellite-year in the INTELSAT V series, we assume that the same ratio of capital costs per satellite-year to payment to capital per satellite-year holds for that series as for the three earlier series, that is, a ratio of about 4.18. Thus, INTELSAT V payment to capital per satellite-year is set at \$5,555,556 ÷ 4.18 = \$1,329,080.

To complete the compilation of fixed costs per satellite-year for the first three satellite series, a minor technical correction is necessary to assure that all costs are correctly included. For convenience, the following abbreviations will be used:

TABLE 2.5

Capital Costs Per Satellite and Per Satellite-Year
($000, 1965 Prices)

Series	Total Capital Expenditures	Number of Successful Launches	Capital Costs Per Satellite
INTELSAT I-II	52,936	4	13,234
INTELSAT III	83,220	5	16,644
INTELSAT IV	208,304	6	34,717
INTELSAT V	—	—	50,000
Series	**Number of Years in Orbit[a]**	**Capital Cost Per Year Per Satellite**	**Number of Satellite-Years Per Series**
INTELSAT I-II	3	4,411	12
INTELSAT III	5	3,328	25
INTELSAT IV	7	4,960	42
INTELSAT V	9[b]	5,556	—

[a]Accounting lifetime.

[b]Based on mid-range estimate in INTELSAT V Planning Document, September 1973. BG-4-32, Sept. 7, 1973.

Sources: Communications Satellite Corporation, Pocket Guide to the Global Satellite System, March 1974, pp. 12-14.

CAP = capital expenditures
PMTCAP = payment to return on capital
COE = current operating expenses
DEBC = departmental expenses before capitalization

The point of departure is the accounting identity implicit in INTELSAT financial records: total costs = CAP + PMTCAP + COE. It will be recalled that DEBC was chosen to represent short-run variable costs, and that all other costs were assigned to the fixed category. Developing this relation,

$$\text{Total costs} = \text{CAP} + \text{PMTCAP} + \text{COE} + (\text{DEBC} - \text{DEBC})$$
$$= [\text{CAP} + \text{PMTCAP} + (\text{COE} - \text{DEBC})] + \text{DEBC}.$$

Thus, the expression CAP + PMTCAP + (COE - DEBC) represents short-run fixed costs. Its components CAP and PMTCAP have been developed

TABLE 2.6

Payments to Return on Capital, All Series
($000)

Year	Undeflated	1965 Prices, Deflated at 14 Percent Annually
1965	0	0
1966	0	0
1967	-7,667	-5,895
1968	7,670	5,170
1969	9,236	5,468
1970	43,745	22,704
1971	35,620	16,207
1972	34,977	13,990
1973	31,514	12,503
1974	40,002*	12,321
Totals	195,097	82,468

Note: Pro-rata assignment of $82,468,000 to individual series (using percentages from Table 2.4):

INTELSAT I-II	12,700	(15.4%)
INTELSAT III	19,875	(24.1%)
INTELSAT IV	49,893	(60.5%)
Totals	82,468	(100.0%)

*Budget figure.

Sources: Interim Communications Satellite Committee, ICSC/F-30-3, p. 1; INTELSAT Board of Governors, Attachment 2.

above for all four series of INTELSAT satellites and calculated in units of 1965 dollars per satellite-year. The same is now done for the final component of fixed costs, the expression COE - DEBC. Its total in 1965 dollars for the period 1965-74 is estimated at $2,594,000 (rounded to the nearest thousand) from INTELSAT records. The expression is actually negative for the years 1967-69, inclusive. Assigning these costs according to the percentages in Table 2.4, we obtain assignments of the COE - DEBC total as follows (rounded): $400,000 to INTELSAT I-II; $625,000 to INTELSAT III; and $1,569,000 to INTELSAT IV. This yields rounded per satellite-year figures of $33,000 ($400,000 ÷ 12) for INTELSAT I-II; $25,000 ($625,000 ÷ 25) for INTELSAT III; and $37,000 ($1,569,000 ÷ 42) for INTELSAT IV. Each of these figures is approximately 1/133 of the total capital costs per satellite-year in the respective series. Assuming the same ratio for INTELSAT V, we obtain a COE - DEBC correction of $42,000 ($5,555,556/133) per satellite-year for that series.

TABLE 2.7

Summation of Fixed Costs, By Series, 1965 Prices
($000 Per Satellite-Year)

Series	Capital Expenses	Payment on Return to Capital	COE less DEBC[a]	Total Fixed Costs
INTELSAT I-II	4,411	1,058	33	5,502
INTELSAT III	3,328	795	25	4,148
INTELSAT IV	4,960	1,188	37	6,185
INTELSAT V	5,556	1,329	42	6,927

[a]Current Operating Expenses Minus Departmental Expenses Before Capitalization

Source: Compiled by the author.

TABLE 2.8

Fixed Cost Per Satellite, and Satellite Capacities

Series	Fixed Cost Per Satellite-Year[a] ($000, 1965)	Nominal Satellite Lifetime (years)	Fixed Cost Per Satellite ($000, 1965) (COST)[b]	Capacity in Units (CAP)
INTELSAT I-II	5,502	3	16,506	480
INTELSAT III	4,148	5	20,740	2,400
INTELSAT IV	6,185	7	43,295	10,000
INTELSAT V	6,927	9	62,343	25,000

[a]Table 2.7

[b]Product of second and third columns.

Sources: Communications Satellite Corporation, Pocket Guide to the Global Satellite System, March 1974, and INTELSAT V System Definition Study, BG-4-32, September 1973.

Now the three components of short-run fixed costs per satellite-year can be summarized for each of the four satellite series (see Table 2.7).

RETURNS TO SCALE IN FIXED (CAPITAL) COSTS

Returns to scale in fixed costs can be estimated with the help of data from Table 2.7. The parameter to be estimated, however, will not be used in the present chapter, but is instead a component of investment cost minimization models developed by the author. Table 2.8 indicates the data needed to estimate that parameter. Using the data in the fourth and fifth columns of Table 2.8, a logarithmic regression of the following specification was run:

$$COST = k(CAP)^{a}.$$

This yielded a value of 0.349 for the parameter a, with a t-value of 5.78 (significant at the 95 percent level). A value of less than unity for parameter a indicates increasing returns to scale in capital expenditures, and a value as low as 0.349 shows that they are marked. Such economies are of a dynamic nature reflecting technological change as well as scale economies in the usual static sense. It should be noted that the regression performed has only two degrees of freedom, and as such is more of a rough-and-ready engineering estimate than a true statistical estimation.

TOTAL, AVERAGE, AND MARGINAL SHORT-RUN COSTS

The total, average, and marginal short-run cost functions for the individual series can now be derived. Using Table 2.7 for fixed costs and equation 2.1 for variable costs, we obtain the following relations for short-run total costs (SRTC), expressed in thousands of 1965 dollars per satellite-year for the satellite series indicated by the subscript:

$$SRTC_{I-III} = 5,502 + 1,440(TRAF)^{0.22}$$
$$SRTC_{III} = 4,148 + 1,440(TRAF)^{0.22}$$
$$SRTC_{IV} = 6,185 + 1,440(TRAF)^{0.22}$$
$$SRTC_{V} = 6,927 + 1,440(TRAF)^{0.22} \qquad (2.2)$$

Short-run average costs (SRAC) are obtained by dividing equation 2.2 by the level of traffic (TRAF) for each series:

$$
\begin{aligned}
SRAC_{I-II} &= 5,502 \div TRAF + 1,440(TRAF)^{-0.78} \\
SRAC_{III} &= 4,148 \div TRAF + 1,440(TRAF)^{-0.78} \\
SRAC_{IV} &= 6,185 \div TRAF + 1,440(TRAF)^{-0.78} \\
SRAC_{V} &= 6,927 \div TRAF + 1,440(TRAF)^{-0.78}
\end{aligned}
\quad (2.3)
$$

The short-run marginal cost (SRMC) of the last unit of traffic is obtained by differentiating equation 2.2 with respect to the traffic level:

$$
\begin{aligned}
SRMC_{I-II} = SRMC_{III} = SRMC_{IV} = SRMC_{V} &= 0.22 \times 1440(TRAF)^{-0.78} \\
&= 317(TRAF)^{-0.78}
\end{aligned}
\quad (2.4)
$$

Equation 2.4 is a reminder that marginal costs are the same for all satellite series. This is the case because the variable cost function is the same for all series.

LONG-RUN COSTS FOR SINGLE-SATELLITE CONFIGURATIONS

The relations given in equations 2.2, 2.3, and 2.4 are still somewhat artificial in that they express costs only for a single satellite, whereas at any given moment INTELSAT may have several operating and standby satellites in orbit over each of the three ocean regions. It is of theoretical interest, however, to attempt to construct a long-run average cost envelope curve for the four single-satellite SRAC curves in equation 2.3. Subsequently, the more realistic situation of several satellites from each series will be treated, and cost curves will be derived and applied to actual traffic experience.

According to the textbook paradigm, the long-run average cost curve is the lower envelope of the short-run average cost curves, containing the locus of points at which short-run configurations of productive capacity are optimally adjusted to factor prices. It is tangent to a point on each U-shaped SRAC curve and is itself U-shaped.

The paradigm is of little use in a decreasing-cost technology such as satellite communications, since U-shaped average cost curves contain ranges of increasing cost beyond their lowest point. But the SRAC curves expressed in equation 2.3 are defined only up to the point of maximum capacity in each series; at those points, as well as all other positive traffic values at less than maximum capacity, the SRAC curves will be downward sloping. Hence, their points of lowest average cost will be at the maximum capacities indicated in Table 2.8, column 5. These maximum-capacity average costs are given below.

$$\begin{aligned}
SRAC_{I-II}(480) &= 5,502 \div 480 + 1,440(480)^{-0.78} = \$23,141 \\
SRAC_{III}(2,400) &= 4,148 \div 2,400 + 1,440(2,400)^{-0.78} = \$5,558 \\
SRAC_{IV}(10,000) &= 6,185 \div 10,000 + 1,440(10,000)^{-0.78} = \$1,711 \\
SRAC_{V}(25,000) &= 6,927 \div 25,000 + 1,440(25,000)^{-0.78} = \$812
\end{aligned} \tag{2.5}$$

The SRAC curves for higher capacity satellites have substantially the same average costs as the other curves do at low capacity levels. Their main difference, as the curves in equation 2.5 indicate, is that some extend farther along the horizontal axis than do others. This means that they are capable of sustaining those essentially equal declining unit costs at higher capacities. This fact sheds light on the nature of technological progress occurring between successive series of INTELSAT satellites. One characteristic of that progress, highlighted in equation 2.3, is the pushing back of scale frontiers in a decreasing cost industry. Another, which has contributed to the near-equality of average costs even at low traffic levels, is the increasing satellite lifetime from series to series. Surely, technological progress occurs when a given productive unit is identical to an earlier one in all particulars except in lasting twice as long. It was for this reason that care was taken above in measuring satellite outputs and costs in terms of satellite-years and unit-years. If costs were measured per satellite, without a time dimension, the later series, having a higher capital cost per satellite, would have higher average costs over low traffic ranges than would the earlier satellites. But the increased lifespan of the later series (borne out in observation) has compensated for the higher per-satellite costs and brought average costs per satellite-year to near equality. As has been argued above, decreases in short-run average costs can be attributed either to dynamic technological progress or to static scale economies. Yet equations 2.2 and 2.3 show clearly that the extent of scale economies (reduction of unit costs with increasing output) is very similar for all four series, with no such shifts evident. The progress involved, then, is not in increasing the rate at which those economies are realized from series to series but rather in extending the scale up to which those (similar) economies can be realized. Because of this, one avoids the intertemporal difficulties which might otherwise arise in comparing different short-run cost curves and in using them to derive long-run curves, as will be done below.

In constructing a long-run average cost (LRAC) curve from the four SRAC curves given in equation 2.3, an expression is needed with no term for average fixed costs, since there are no fixed costs in the long run. The relation is therefore specified as

$$LRAC = k(TRAF)^{a}, \tag{2.6}$$

where presumably $0 \leq a \leq 1$, indica(ting) decreasing long-run average costs over the entire range of output. A problem arises in selecting points to which equation 2.6 is to be fit. (For a family of U-shaped SRAC curves, the LRAC curve, as noted, is an envelope from below which is tangent to each SRAC and which is itself U-shaped.) Obviously, no such curve can be fit to the SRAC curves given in equation 2.3. The $SRAC_{I-II}$ curve, for example, lies entirely northeast of the $SRAC_{III}$ curve for all levels of its output, so tangency of a curve of the form specified in equation 2.6 to all four of those SRAC curves is a mathematical impossibility. Furthermore, tangency of equation 2.6 to the remaining curves, $SRAC_{III}$, $SRAC_{IV}$, and $SRAC_{V}$, is by no means assured since they do not "bottom out" or reach a mimimum level reflecting constant returns, but rather decline up to their levels of full capacity. Therefore, equation 2.6 will be fit through the points of maximum capacity of those three curves, that is, the capacity-SRAC pairs (2,400 and 5,558), (10,000 and 1,711), and (25,000 and 812) given by equation 2.5. To add another degree of freedom, a point is added at the capacity level of 480 units, the maximum capacity of the INTELSAT I-II series. The cost coordinate chosen for that value is not the INTELSAT I-II SRAC ($23,141), but the INTELSAT III SRAC at that traffic level, which is $20,312 and lower than that of the other series at that level. Fit through these four points, the log-linear least-squares regression of equation 2.6 yields the relation

$$\underset{}{LRAC} = \underset{(583.18)}{3,140,687}\underset{(-268.85)}{(TRAF)^{-0.81}} \tag{2.7}$$

where the t-values are indicated in parentheses and costs are in 1965 dollars. The corresponding total and marginal long-run cost curves are

$$LRTC = 3,140,687(TRAF)^{0.19}$$

and

$$LRMC = 596,731(TRAF)^{-0.81} = 0.19 \times LRAC.$$

The theoretical nature of equation 2.7 will again be emphasized. First, it represents only what costs would be at given capacities if all factors of production were continuously variable, that is, if there were no fixed or sunk costs. This condition is notably lacking in the case of INTELSAT satellites, for which there is considerable lumpiness in productive capacities. Second, equation 2.7 is theoretical in that it is an envelope of SRAC curves for individual satellites rather than for operating configurations of several satellites. Attention will now be focused on the task of aggregating cost curves for several satellites to approximate past INTELSAT operating experience.

DERIVATION OF AGGREGATE COST CURVES

Suppose it is known that at a particular moment there are n_1 satellites of series I-II, n_3 satellites of series III, and n_4 satellites of series IV which have been successfully launched and are either operational or on standby status. How can the aggregate cost curves of such a configuration of satellites be derived from the single-satellite curves summarized in equations 2.2, 2.3, and 2.4?

To begin with, aggregate costs will be conceptually divided into fixed and variable costs in the short run, as previously. Variable costs should pose no problem; they are incurred on the ground as primarily labor and other administrative costs and are not peculiar to any individual satellite or satellite series. Thus, if the total traffic level carried over a particular configuration of different satellites is TRAF, annual total variable costs are, using equation 2.1,

$$1,440,000(\mathrm{TRAF})^{0.22},$$

as they were seen to be for traffic carried over individual satellites.

For a given group of in-orbit satellites, fixed costs are derived from the same sources as before: capital expenditures, payments to return on capital, and the technical correction to prevent double counting of costs. Yet a decision must be made regarding the way in which these costs are to be aggregated both within and among satellite series. For example, do scale economies exist between satellites of a given series? If the first satellite procured from a contractor costs a certain amount, does the second, identical in all features, cost less? Furthermore, is there an interseries externality pattern whereby procurement of satellites in a given series reduces later procurement costs of another series, particularly if both are obtained from the same contractor? It cannot be denied a priori that such nonlinearities may exist; indeed, the procurement of initial INTELSAT IV satellites from Hughes Aircraft provided for quite gradually decreasing marginal costs of later satellites in that series, but nowhere near the extent of the scale economies observed from series to series. Furthermore, interseries externalities, to the extent they exist, have already been captured in the single-satellite cost equations. Remaining nonlinearities in fixed costs may exist but will be assumed at this point as being qualitatively small and statistically unobtainable to the extent they are present at all.

Fixed costs will therefore be combined linearly in the simplest fashion. Let SAT12, SAT13, and SAT4 represent the number of satellites of each of those series in orbit and undepreciated at a given moment. Then aggregate total fixed costs for that particular configuration will be, in thousands of 1965 dollars per year, using equation 2.2,

$$5{,}502 \times SAT12 + 4{,}148 \times SAT3 + 6{,}185 \times SAT4.$$

Adding this expression to the variable costs noted above, we obtain aggregate short-run total costs (AgSRTC) as follows:

$$AgSRTC = 5{,}502 \times SAT12 + 4{,}148 \times SAT3 + 6{,}185 \times SAT4 + 1{,}440(TRAF)^{0.22} \quad (2.8)$$

costs being expressed as usual in thousands of 1965 dollars per year. Aggregate short-run average costs (AgSRAC) are then

$$AgSRAC = \frac{5{,}502 \times SAT12 + 4{,}148 \times SAT3 + 6{,}185 \times SAT4}{TRAF} + 1{,}440(TRAF)^{-0.78} \quad (2.9)$$

and the (marginal) cost of the last unit of traffic is

$$AgSRMC = 0.22 \times 1{,}440(TRAF)^{-0.78} = 317(TRAF)^{-0.78} \quad (2.10)$$

as before.

Aggregate short-run cost curves will now be applied to detailed past operating experience (note that costs of the planned INTELSAT V series are not included in the equations). The data selected include the in-orbit satellite configuration, the equivalent full-time traffic level (including television and other part-time usage converted to full-time units), and the INTELSAT space segment charge per unit-year in force for each of the 114 months from July 1965 through December 1974, inclusive. For each month, equations 2.8, 2.9, and 2.10 were used to derive aggregate short-run total, average, average variable, and marginal costs. Figure 2.1 indicates twice-yearly abstractions from some of these data, showing AgSRAC, average variable costs, and the level of INTELSAT's basic space segment utilization charge.

Since all costs are expressed in 1965 dollars according to some discounting pattern, a discount rate had to be chosen for INTELSAT's basic space segment utilization charge, which is essentially a measurement of INTELSAT's average costs over a long period and will be compared below with various cost and revenue figures. Unfortunately, there is no document to which we can turn to find a statistic such as "average cost of one unit of satellite utilization" to help select a

FIGURE 2.1

Aggregate Average and Average Variable Cost, and Annual Space Segment Charge, July 1965 through December 1974

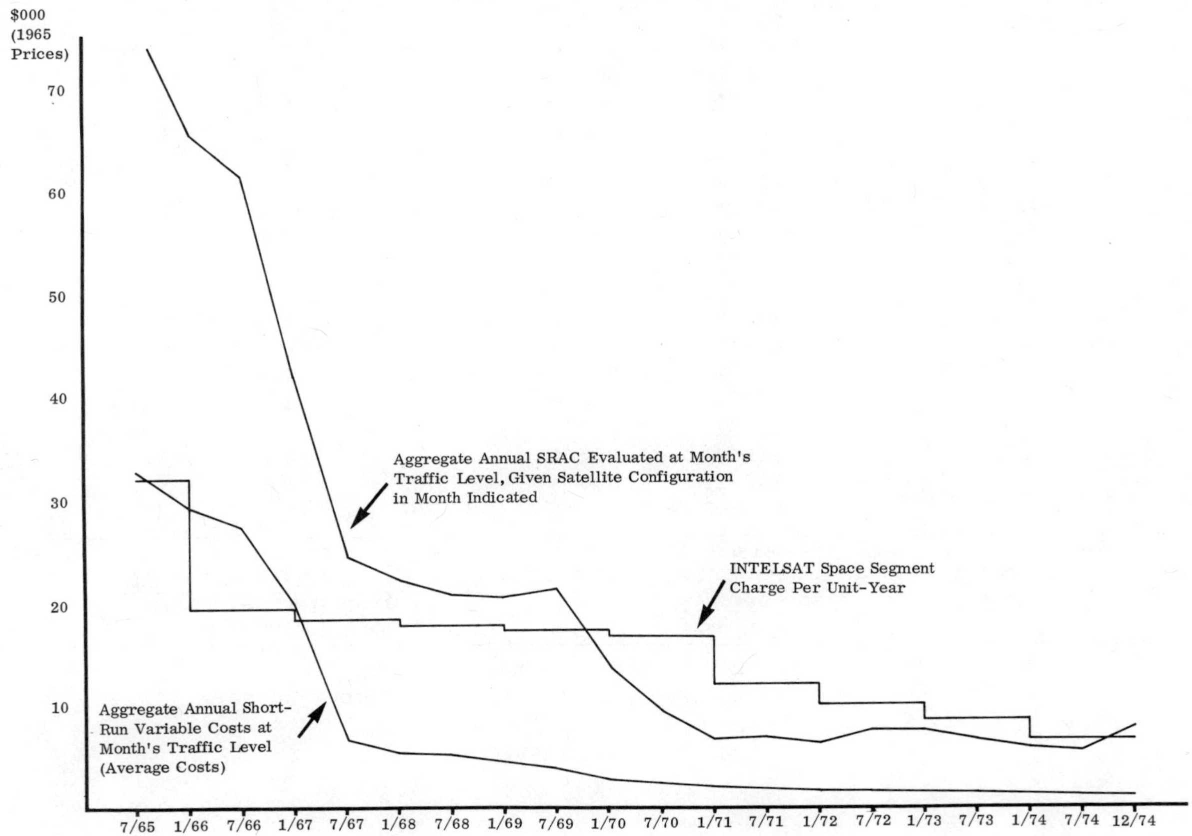

TABLE 2.9

Satellite Configurations in Use, 1965-74, with Capacity and Short-Run Average Capacity Costs

Configuration	Number of Satellites From Series Indicated			Capacity (units)	Short-Run Average Costs at Capacity (1965 Prices)	Month and Year Configuration First Employed
	INTELSAT I-II	INTELSAT III	INTELSAT IV			
1	1	0	0	480	23,130	7/65
2	2	0	0	960	18,258	2/67
3	3	0	0	1,440	16,415	4/67
4	4	0	0	1,920	15,420	11/67
5	4	1	0	4,320	8,157	1/69
6	4	2	0	6,720	5,999	3/69
7	4	3	0	9,120	4,951	6/69
8	3	4	0	11,040	4,009	2/70
9	1	4	0	10,080	3,227	4/70
10	1	5	0	12,480	3,022	5/70
11	0	5	0	12,000	2,676	11/70
12	0	5	1	22,000	1,814	4/71
13	0	5	3	42,000	1,292	3/72
14	0	5	4	52,000	1,177	8/72
15	0	5	5	62,000	1,097	9/73
16	0	4	5	59,600	1,069	1/74
17	0	3	5	57,200	1,039	3/74
18	0	2	5	54,800	1,006	8/74
19	0	2	6	64,800	955	12/74

Source: Communications Satellite Corporation, Pocket Guide to the Global Satellite System, March 1974, pp. 12-14, and updated through December 1974. Retirements from in-orbit service based on failure data and accounting lives.

suitable discount rate. INTELSAT is the only organization offering such services, and its steadily declining space segment charge reflects technological advances much more strongly that it does any evolution in the general price structure. For this reason, the discount pattern was selected as a weighted composite of the three discount factors already applied to INTELSAT costs, under the reasoning that it is precisely those costs which INTELSAT's utilization charge is designed to recover. Previously, capital costs were deflated by an index of the cost of capital equipment in the electrical equipment industry; payments to return on capital were deflated at the rate of 14 percent annually; and variable costs were deflated at a rate reflecting the evolution of the wage rate in the electrical equipment industry. A summation indicates that during the 9-year period 1964–72, those three cost categories accounted, respectively, for 72.2 percent, 13.0 percent, and 14.8 percent of total INTELSAT costs, in 1965 dollars. Thus, for each of the 114 months for which operating data were assembled, those percentages were used to weight the three separate discount factors mentioned. To approximate procedures used previously, the discount rate is applied annually rather than monthly. As an alternative, annual discount rates of 2 percent, 4 percent, 6 percent, and 8 percent were applied to the INTELSAT utilization charge, but the composite discount rate derived by the procedure above, which is not far from a 4 percent annual discount rate, was found to be satisfactory, as will be seen below.

The aggregate long-run cost curves will now be derived in a fashion analogous to that used in the single-satellite abstraction above. Specifically, the long-run average cost envelope will be fitted, as before, through the traffic—average cost pairs corresponding to the maximum capacities of the different satellite configurations historically employed by INTELSAT. There were nineteen different satellite configurations observed in the 114-month period being examined, and those configurations, together with the corresponding capacity and short-run average cost at full capacity, are presented in Table 2.9.

So that the aggregate LRAC curve will correspond as closely as possible to a lower envelope of the aggregate SRAC curves, all those capacity/cost-at-capacity pairs that are "northeast" of any other such pairs will be eliminated. In Table 2.9, configurations 8, 10, 15, 16, and 17 represent such pairs and were removed from the observations. The remaining 14 pairs were used to calculate coefficients of the aggregate long-run average cost (AgLRAC) curve, which is specified as follows:

$$\text{AgLRAC} = k(\text{TRAF})^{a}.$$

A least-squares log-linear regression yielded the following relationship when the 14 points described above were used:

$$AgLRAC = \underset{(51.03)}{2,649,728}(TRAF)^{\underset{(-22.55)}{-0.71}} \quad R^2 = .9769 \tag{2.11}$$

As might be expected, equation 2.11 shows less marked decreasing costs than does equation 2.7, the LRAC curve for single satellites. From equation 2.11 the following aggregate, long-run total and marginal cost curves are derived:

$$AgLRTC = 2,649,728(TRAF)^{0.29} \tag{2.12}$$

$$AgLRMC = 768,421(TRAF)^{-0.71} = 0.29 \times AgLRAC \tag{2.13}$$

NOTES

1. For short-run average cost functions of this shape, based on Cobb-Douglas technology assumptions, see Theodore E. Keeler, "Railroad Costs, Returns to Scale, and Excess Capacity," Review of Economics and Statistics 56, no. 2 (May 1974), pp. 201-08, and G. Eads, M. Nerlove, and W. Raduchel "A Long-Run Cost Function for the Local Service Airline Industry: An Experiment in Nonlinear Estimation," Review of Economics and Statistics 51, no. 3 (August 1969), pp. 258-270.

2. For a discussion of the static and dynamic aspects of apparent scale economies over time in satellite communications, see Marcellus S. Snow, "Investment Cost Minimization for Communications Satellite Capacity: Refinement and Application of the Chenery-Manne-Srinivasan Model," Bell Journal of Economics 6, no. 2 (Autumn 1975). This study, however, deals with capacity investment costs rather than with variable costs.

CHAPTER 3
A CONSIDERATION OF MARGINAL COST PRICING IN THE FRAMEWORK OF THE INTELSAT TARIFF

With some important exceptions to be noted later, INTELSAT, like most regulated public utilities, has few tariffed service offerings that come close to reflecting the marginal cost pricing principles one finds defended in textbooks on microeconomic theory and public utility economics. This divergence between theory and reality can be traced to two major sources. The first deals with the level of economic knowledge and sophistication brought to bear in the process of making and enforcing rates for public utility services. While economists, financial analysts for larger firms, and perhaps an occasional judge or official on a regulatory board might be familiar with the theoretical issues involved in marginal cost pricing, the awareness and debate has by and large been confined to books, journals, and college classrooms and has bypassed the corridors of power.

The second and perhaps more important reason for the lack of marginal cost pricing observed in public utility tariffs derives not from ignorance of the theory and issues in question (economists are perhaps too often self-righteous on this count), but from the quite considerable theoretical and practical shortcomings of different approximations to marginal cost pricing. These difficulties are familiar both to academic economists and economists and others involved in the practical construction and evaluation of charging systems for public utilities. The following section will examine the effect that marginal cost pricing might have on INTELSAT's finances; theoretical difficulties in marginal cost pricing; and methods proposed to recover operating deficits caused by marginal cost pricing. At this point in the discussion, the simplifying assumption will be maintained that INTELSAT sells a single service, the unit of utilization, for periods of one year or longer.

INTELSAT AND MARGINAL COST PRICING

In the presence of decreasing costs, reflected by a downward-sloping average cost curve, marginal costs are less than average costs, so that a tariff which charged marginal costs for all units sold would result in an operating deficit. As will be seen, INTELSAT is no exception. Its strongly decreasing costs, indeed, prevent all but a small fraction of operating costs from being recovered by marginal cost pricing.

In the INTELSAT cost model developed previously a number of different interpretations might be given to "marginal cost" charging. Some are extreme. Alfred Kahn, for example, cites (and rejects as impractical) a scenario in which a potential passenger boards an aircraft with some empty seats and its engines warmed up, ready for takeoff. In the shortest run, the marginal costs of carrying this late arrival are essentially zero, aside from the added fuel his weight might require. Yet if the flight were full and the passenger had to ride a supplementary aircraft alone, he would be causally responsible for all expenses of the flight and would have to pay a small fortune as "his" marginal cost. [1] Likewise, an INTELSAT customer might find that just as his traffic is about to be carried, the satellite for which it was intended becomes filled, and a new one must be brought in just for him, entailing enormous "short-run" marginal costs. For reasons of equity and practicality, Kahn and virtually all other writers reject this extreme short-range interpretation of marginal cost pricing. In its rejection, the idea of equity prevails: the potential passenger has no control over whether he is the last passenger on a full plane or the only passenger on an empty one. It seems fairer to average the marginal costs over all passengers, apart from considerations of averaging all costs in the rate structure. This is not, of course, a retreat from peak-load pricing, designed to discourage usage at times over which the consumer presumably does exercise some control. [2]

This averaging of marginal costs is evident in the cost equations for INTELSAT derived in Chapter 2. Indeed, there is an implicit abstraction from any assignment of peak-loading cost causation of individual units of traffic; all such capacity costs are included in fixed costs and averaged over all users rather than being included in variable and marginal costs, as they rightly should be in a system with periodic excess demand. Now, for a number of reasons, [3] supply has outstripped demand to such an extent during the first decade of INTELSAT's operation that no problem resembling peak-loading congestion, with its accompanying presumption of causal assignment of capacity costs to demand occurring at peak traffic times, has ever occurred, nor does one appear likely to arise in the future. This leaves, therefore, validly assignable marginal (variable) costs of the type described, namely, on-ground departmental expenses before capitalization (DEBC) which were shown to differ significantly with the traffic level but not with the level of capacity.

In determining the level of INTELSAT revenues that could have been generated under various tariff structures during the 114-month period from mid-1965 through the end of 1974, attention will be focused only on the aggregate short-run equations 2.8, 2.9, and 2.10 and on the accompanying aggregate long-run equations 2.11, 2.12, and 2.13 rather than on the earlier single-satellite equations, since it is the short-run equations that approximate actual INTELSAT experience and orbital configurations. Furthermore, economic theory requires that the marginal cost of the last unit sold be equated to price in order for the usual optimality properties to apply. Earlier units are, properly speaking, inframarginal and do not measure opportunity costs at the margin. Therefore, all following discussion of marginal cost charging will focus exclusively on the marginal cost of the last unit sold as the basis of marginal cost charging.

What proportion of INTELSAT costs can or could have been recovered by a marginal cost charging policy, assuming aggregate cost functions of the types indicated by equations 2.8 through 2.13? This can be answered on the basis of those cost functions. For short-run costs, where aggregate short-run average variable cost is given by

$$\text{AgSRAVC} = 1{,}440(\text{TRAF})^{-0.78}$$

the relationship between aggregate average cost (equation 2.9) and aggregate marginal cost (equation 2.10) is given by

$$\text{AgSRMC} = 0.22 \times \text{AgSRAVC}.$$

The factor 0.22 is, of course, the exponent of the traffic term indicating decreasing costs and originally derived in equation 2.1. The relation above indicates that only 22 percent of short-run variable costs can be recovered by setting price equal to the cost of the last or marginal unit sold. The percentage of all short-run costs (fixed as well as variable) recoverable by marginal-cost charging will of course be even less and will depend on the level of fixed costs.

For long-run aggregate costs, with no fixed-cost term, the relationship between equations 2.11 and 2.13 is even more straightforward and has already been expressed:

$$\text{AgLRMC} = 0.29 \times \text{AgLRAC}.$$

Thus, marginal-cost charging would recover 29 percent of long-run costs; the factor 0.29 is the exponent of the long-run average cost envelope curve derived in Chapter 2.

These relationships can be illustrated by using data for one of the months from the 114-month sample, say, September 1971. From equation 2.8, aggregate short-run total cost for a year at that month's traffic level is calculated to be

$$5,502 \times (0) + 4,148 \times (5) + 6,185 \times (1) + 1,469 \times (5,407)^{0.22}$$

in thousands of 1965 dollars, or $36,466,831 on an annual basis, where traffic during that month was 5,407 units on an annual basis. Assuming linearity of total costs when divided into periods of less than a year, that makes total costs of $3,038,903 for September 1971. The average annualized cost at the traffic level of 5,407 units annually is $6,744 per unit. Of total annual costs, variable costs are as follows: $1,440,000(5407)^{0.22}$ = $9,541,440, or 26.2 percent; the remainder (73.8 percent) are fixed. As noted, short-run application of marginal-cost charging would yield 22 percent of short-run variable costs, or $2,099,117, which, as a portion of total costs, is only 0.22×0.262 = 5.8 percent.

Generally, variable costs ranged from about 44 percent of total costs in the early months of INTELSAT operations to about 20 percent of total costs in late 1974. Revenues yielded by marginal cost charging would therefore have yielded between about 10 percent and 4 percent of total short-run costs during that period. Table 3.1 summarizes this information.

RECOVERY OF TOTAL COSTS WITHIN A MARGINAL PRICING FRAMEWORK

To salvage the marginal cost pricing concept in the face of decreasing costs, many writers have suggested methods by which revenue derived from marginal cost charging might be supplemented to meet total operating costs. Three primary methods suggested are (1) the addition of a fixed percentage to all marginal costs that is high enough to recover operating expenses; (2) government subsidization of the decreasing-cost industry to make up the operating deficit; and (3) some form of price discrimination.

The first method has been attempted by Electricité de France in recovering its operating deficit.[4] Early writers in the field, such as Harold Hotelling, Ragnar Frisch, and James Meade,[5] asserted that marginal conditions for welfare maximization were not upset if a proportional increment were added to all marginal costs. Later writers, however, including Abba Lerner and Paul Samuelson, showed that this was not the case. Their refutation went along these lines:

> . . . if the prices of consumers' goods were not equal to marginal cost, but factor payments were maintained equal to the marginal products of the factors, the relationship

TABLE 3.1

Effects of Marginal Cost Pricing on INTELSAT
($000, 1965 Prices)

Period	Short-Run Fixed Costs	Short-Run Variable Costs	Short-Run Total Costs	Percent of Short-Run Variable Costs Recovered By Marginal Cost Charging	Percent of Short-Run Total Costs Recovered By Marginal Cost Charging
September (1971) (annualized)	26,926 (73.8% of SRTC)	9,541 (26.2% of SRTC)	36,467	22.0	5.8
June 1965-December 1974	253,885 (75.2% of SRTC)	83,826 (24.8% of SRTC)	337,711	22.0	5.5

Source: Compiled by the author.

between work and leisure would be altered and the marginal conditions throughout the system would not be met. On the other hand, if prices of factors as well as finished goods were all raised proportionally, nothing would be accomplished; it would still be found that prices were equal to marginal costs.[6]

A second, more practical difficulty presents itself in the case of a decreasing-cost industry in which output is viewed as homogeneous. If marginal cost covers, say, x percent of average cost (x being less than 100), and price is increased by a factor of 100/x over cost so that the operating deficit is covered, all one has done is converted marginal cost into average cost. Thus, only in industries with nonequal marginal costs for two or more distinct outputs will an equal percentage increment of marginal cost recover an operating deficit and at the same time avoid equality with average cost.

A second method of supplementing revenues provided by marginal-cost charging in the presence of decreasing costs involves government subsidization of one form or another. The extent to which welfare is still maximized after marginal costs are supplemented depends crucially on the way in which the government raises the subsidy. Hotelling originally argued that deficits could best be made up out of government subsidies raised by lump-sum taxes, and he included income, inheritance, and land taxes in that category.[7] Subsequent debate centered on the extent to which any tax, particularly an income tax, could be viewed as a lump-sum tax in the definitional sense that it "falls on either producers' or consumers' surplus, and therefore does not violate the marginal conditions."[8] Use of the income tax to subsidize decreasing cost industries, routinely proposed by early writers on marginal-cost pricing, was called into question later by a number of scholars. Meade and Samuelson, for example, both maintained that income taxes are in effect excise taxes on the supply of factors of production, such as one's own labor, and distort the existing marginal conditions regarding one's preferences toward work and leisure.[9] H. P. Wald went a step further, pointing out that in some cases, depending on the relative elasticities of demand, an excise tax on items for which demand was inelastic might be less distortive of marginal conditions than the income tax, which he viewed as "an excise tax on work."[10] And then Lerner, noting that lump-sum taxes are "impractical or even impossible," and that some final outputs, such as leisure, undeclared mutual services, and household activities, are untaxable,[11] developed taxation rules of a "quasi-optimal" or second-best nature to minimize the "social change" due to these constraints. The first such rule he proposed is that all taxable outputs be taxed such that the ratio of their tax (price less marginal cost) to their marginal cost be equal to each other. The second rule of ideal taxation is that the ratio of the tax to marginal tax revenue be equalized for all outputs, or, in other words, that tax elasticities be equalized. Lerner summarized both rules as follows: "Let the tax on each use of resources be equal to the tax on the alternative use."[12] The effect of these rules, Lerner claims, is that "there will . . . be no incentive to shift any resources to another (and inferior) use so as to avoid tax"; since this would not merely "minimize" any social damage from taxation but would "eliminate it," the outcome, he argues, would be "Pareto optimal."[13]

The search for the perfect nondistorting lump-sum tax to subsidize declining-cost industries has not been resolved and has an ironic twist to it. Eli W. Clemens characterized the difficulty by objecting to

> the easy reliance on the tax system to adjust divergences between average cost and marginal cost despite the fact that the tax system itself is the greatest example of nonmarginal cost pricing.[14]

In other words, if the tax system, blithely accepted by most economists as an engine of income redistribution, is overwhelmingly nonmarginal in its operation, why split hairs over the fine tuning of the marginal cost pricing mechanism and then use the tax system to make up the operating deficit?

This leads to a final objection which will be raised in regard to tax-supplemented operating deficits in decreasing-cost industries.

Clemens characterizes it as "[the] tacit assumption that utility is measurable and that interpersonal comparisons of utility are feasible."[15] Subsidization of a service to any extent by others who do not consume that service in the same proportion as they pay its costs implies a redistribution of income with obvious welfare-theory implications. Consider, for example, what the consequences would be if individual member governments of INTELSAT decided to make up their shares of the INTELSAT operating deficit due to marginal cost pricing by domestic taxation of low-elasticity luxury goods, of inheritances, of land, etc. Even if such taxes were nondistortive of the conditions on which marginal cost pricing depends for its optimality properties, a redistribution of income would have been effected from certain classes of wealthy consumers, heirs, and landowners to consumers of international telephone service, data service, television programs, etc. No writer has yet proposed a way in which the welfare implications of any such redistribution could be settled without implicit interpersonal comparisons of measured utility. Failing such a demonstration, it is clear that noneconomic value judgments must play a role in the application of any marginal-cost subsidy scheme effected by government taxation. Nancy Ruggles suggests heroically that this difficulty can be skirted if the "marginal utility of income is the same for all people,"[16] a highly unlikely possibility. She does, however, provide a useful review of the welfare aspects of marginal cost pricing in a declining-cost industry:

> In order to make the marginal cost pricing system fulful the requirements claimed for it, a method of financing would have to be provided that would meet the marginal conditions. . . . Introducing the compensation would in effect mean that the revenue for subsidizing any given product would have to be derived from the people who consumed the product, and not from anyone else. To do this without violating the marginal conditions, the tax would have to fall on the consumers' surplus derived by the purchasers from the marginal unit purchased by the consumption of that specific product. It could not bear upon the marginal unit purchased by any consumer, so any form of per-unit tax would be inadmissible. A tax that must fall upon the consumption of a specific product, but not upon the marginal unit, would of necessity yield a form of price discrimination. . . . In order, therefore, to reach a pricing system which does actually meet the conditions which the advocates of marginal cost pricing claim their system meets, it would be necessary to abandon marginal cost pricing altogether, and adopt in its stead a special form of price discrimination.[17]

Ruggles noted further:

> . . . it would seem to be true . . . that it is not possible to design a price system upon the basis of some criterion of efficiency, and then to alter the income distribution in any desired way without affecting the efficiency of the pricing system. No such separation of the problem is possible. Every pricing system results in some sort of income distribution, and no substantial redistribution of income is possible without changing that pricing system. For this reason, it is imperative that the question of interpersonal valuations be taken specifically into account. It is because of the failure to recognize this necessity, in maintaining that no interpersonal comparisons are involved, that objections can be raised to the marginal cost pricing thesis. [18]

The reference to price discrimination serves to introduce a study by William J. Baumol and David F. Bradford arguing that price elasticity of demand be used as a basis of recovering operating deficits due to marginal cost pricing. [19] In its simplest form, in the absence of cross-elasticities of demand, the rule they propose

> requires that each price be set so that its percentage deviation from marginal cost is inversely proportionate to the item's price elasticity of demand. According to this result, the social welfare will be served most effectively not by setting prices equal or even proportionate to marginal costs, but by causing unequal deviations in which items with elastic demands are priced at levels close to their marginal costs. The prices of items whose demands are inelastic diverge from their marginal costs by relatively wider margins. [20]

Baumol and Bradford admit that this result is "curious" and nonintuitive, as it seems to imply that ordinary price discrimination might come close to maximizing social welfare in the presence of a profit constraint. [21] Thus, like Lerner in the taxation case, the authors claim to have found a true second-best solution under the constraint that marginal cost pricing alone does not cover operating costs:

> In a world in which marginal cost pricing without excise or income taxes is normally not feasible, the solution we have usually considered to characterize the "best" is none too good, because it is simply unattainable. In that case, the systematic deviations between prices and marginal costs that [this article] calls for may truly be optimal because they constitute the best we can do within the limitations imposed by normal economic circumstances. [22]

Chapter 5 will present an example of price discrimination offering advantages of this nature to INTELSAT's financial structure. In addition, Chapter 4 will survey the INTELSAT tariff structure and disclose occasional areas of price discrimination already present in INTELSAT's tariff schedule.

The first two methods discussed above as means of recovering the operating deficit, however, would appear impractical for INTELSAT. The equiproportional increase in marginal costs as a basis of pricing reduces to average cost pricing in the case of one output, as has been seen, and its optimality properties are dubious at best. More refined forms of taxation or government subsidy, involving net transfers from nonusers of satellites to users of satellites in order to make up the deficit, involve difficult distributional problems and interpersonal utility comparisons and would be politically infeasible as well. It is inconceivable that INTELSAT members, numbering now nearly ninety, would be able to persuade their fiscal authorities at home to subsidize international telecommunications with tax monies that would have to be diverted from other uses such as welfare, education, public health, and so forth. Indeed, the subsidy in practice is usually in the opposite direction.

NOTES

1. Alfred E. Kahn, The Economics of Regulation: Principles and Institutions, vol. 1, Principles (New York: John Wiley, 1970), pp. 75-76.
2. For an early treatment of peak-load pricing in INTELSAT, based on peak hourly demand of the developed regions, and using future traffic and cost projections for the most part, see Edward Joseph Heiden, "Some Economic Problems of International Space Telecommunications" (Ph.D. diss., Washington University, 1967).
3. See Marcellus S. Snow, "Investment Cost Minimization for Communications Satellite Capacity: Refinement and Application of the Chenery-Manne-Srinivasan Model," Bell Journal of Economics 6, no. 2 (Autumn 1975).
4. James R. Nelson, "Practical Applications of Marginal Cost Pricing," American Economic Review 53, no. 2, (May 1963), p. 480.
5. See Nancy Ruggles, "Recent Developments in the Theory of Marginal Cost Pricing," Review of Economic Studies 17 (1949-50), reprinted in R. Turvey, ed., Public Enterprise: Selected Readings (Baltimore: Penguin Books, 1968), p. 17.
6. Ibid., pp. 17-18. Emphasis in original.
7. Harold Hotelling, "The General Welfare in Relation to Problems of Taxation and of Railway and Utility Rates," Econometrica 6 (1938), pp. 242-69.
8. Ruggles, op. cit., p. 32.
9. Ibid.

10. H. P. Wald, "The Classical Indictment of Indirect Taxation," Quarterly Journal of Economics 59 (1945), pp. 577-96.

11. Abba P. Lerner, "On Optimal Taxes With an Untaxable Sector," American Economic Review 60, no. 3 (July 1970), pp. 284-94.

12. Ibid., p. 285.

13. Ibid., pp. 285-86.

14. Eli W. Clemens, "Discussion," American Economic Review 53, no. 2, (May 1963), p. 484. See also Kahn, op. cit., pp. 130-31.

15. Clemens, op. cit., p. 482.

16. Ruggles, op. cit., p. 36.

17. Ibid., pp. 35-36.

18. Ibid., p. 38.

19. William J. Baumol and David F. Bradford, "Optimal Departures from Marginal Cost Pricing," American Economic Review 60, no. 3 (June 1970), pp. 265-83.

20. Ibid., p. 267.

21. Ibid.

22. Ibid., p. 280.

CHAPTER

4

THE INTELSAT RATE STRUCTURE: IMPLICIT CHARGING PHILOSOPHIES

Attention will now be shifted from a discussion of theoretical issues of marginal cost pricing to a detailed consideration of the actual INTELSAT tariff as it has developed and is presently constituted.

AVERAGE COST PRICING: THE POINT OF DEPARTURE

Throughout its first decade of operation (1965-74) INTELSAT relied primarily upon average cost pricing to recover its operating revenue. The previously mentioned INTELSAT Operating Agreement, which entered into force in 1973, lays down the following principles for normal and specialized usage of the space segment:

> The Board of Governors shall specify the units of measurement of INTELSAT space segment utilization . . . and . . . shall establish INTELSAT space segment utilization charges. Such charges shall have the objective of covering the operating, maintenance and administrative costs of INTELSAT, the provision of such operating funds as the Board of Governors may determine to be necessary, the amortization of investment made by Signatories in INTELSAT and compensation for use of capital of Signatories.
>
> . . . The Board of Governors shall establish the charge to be paid for the utilization of [specialized] services . . . and shall take into consideration the costs associated with the provision of the specilized telecommunications services as well as an adequate part of the general and administrative costs of INTELSAT. In doing so, it shall comply with the

> provisions of the Agreement and this Operating Agreement . . . so as to cover fully the costs directly resulting from the design, development, construction, and provision of such separate satellites and associated facilities as well as an adequate part of the general and administrative parts of INTELSAT. [1]

The interim agreements laid down similar requirements, and INTELSAT's decision-making bodies have in general followed average-cost pricing as a matter of policy. The following statement of its Finance Subcommittee is a fair representation of the policy in practice:

> We recommend that the method to be adopted as the norm [for INTELSAT charging policy] should consist of dividing total system costs, or total apportioned costs of the capacity providing the service or facility in question, by the number of units of that service or facility expected to be requested. [2]

The primary vehicle for recovering average costs has been, of course, the space segment utilization charge. It is determined by estimating the "revenue requirement," which consists of capital costs (reflected as depreciation in the bookkeeping sense), return to capital, and current operating expenses. Then traffic is estimated and divided into the revenue requirement to obtain the space segment charge, as the INTELSAT Finance Subcommittee suggested. Table 4.1 summarizes the level of the space segment utilization charge since 1965.

As Figure 2.1 indicated, the composite-discounted space segment charge did not cover short-run average costs in the INTELSAT system for corresponding years until late in 1969, after which time it has been consistently higher than short-run average costs (with the exception of late 1974). A glance at Figure 2.1 might indicate that over its ten years of operation, INTELSAT has not recovered its operating expenses with the space segment charge. But the wide variations in the early years occurred during a period in which traffic and costs were low, while the mild excess of space segment charge over average costs since 1970 has held for increasingly higher traffic levels. In fact, INTELSAT attained a 14 percent cumulative return on capital by 1971 and has held at that level ever since. [3] During the 114-month period from mid-1965 through the end of 1974, total INTELSAT costs in 1965 dollars were estimated to be $337,711,000. If all INTELSAT traffic had paid the full-time space segment utilization charge (discounted by the composite formula) during that period, revenues of $417,531,000, or 23.6 percent greater than total costs, including return to capital, would have been realized. There is, of course, no particular reason these two figures should be equal. First, the discounting of costs and of the space segment charge is to some extent arbitrary; indeed, there must exist a discounting scheme for which the two figures would equal each other. Furthermore, not all INTELSAT traffic paid the full utilization rate or an equivalent rete during the period examined.

TABLE 4.1

INTELSAT's Full-Time Space Segment Charge For a One-Year Allotment of One Unit of Utilization, Variously Discounted to 1965 Prices

	Level of Charge ($)					
	Current	Composite	Annual Discount			
Period	Dollars	Discount[a]	2%	4%	6%	8%
July 1965-December 1965	32,000	32,000	32,000	32,000	32,000	32,000
January 1966-December 1970	20,000	19,280[b] 16,560[c]	19,608[b] 18,115[c]	19,231[b] 16,439[c]	18,868[b] 14,945[c]	18,519[b] 13,612[c]
January 1971-December 1971	15,000	12,060	13,321	11,858	10,575	9,452
January 1972-December 1972	13,000	10,127	11,317	9,879	8,646	7,585
January 1973-December 1973	11,160	8,448	9,525	8,155	7,002	6,029
January 1974-December 1974	9,000	6,615	7,531	6,323	5,327	4,502

[a]See Section A(6) for an explanation of the discounting method used to obtain the composite space segment discount.

[b]Charge level discounted from 1966.

[c]Charge level discounted from 1970.

Sources: INTELSAT, Interim Communications Satellite Committee and Board of Governors meetings, Summary Records.

A survey of the INTELSAT tariff later in this study will disclose a number of divergences, such as preferential rates for volume long-term leasing, cable restoration, and so on. But the fact that total economic costs as calculated in Chapter 2 appear to be close to accounting revenues pegged to recover accounting costs is an independent vindication of the cost derivation in Chapter 2.

In conclusion, it appears indisputable that an average cost philosophy has been the cornerstone of INTELSAT's charging policy since it began operation. Yet there are exceptions to this prevailing philosophy which can be detected in the INTELSAT tariff, in the recommendations of its Finance Subcommittee and of individual member countries, and in planning and background reports it has issued with reference to offering particular services. Attention will now be focused on these areas.

SHORT-TERM SERVICES: METHODOLOGICAL COMMENT

Throughout the derivation of the various cost functions in Chapter 2, the unit of output was assumed to be measured in unit-years. (It will be recalled that a "unit of utilization" has been defined by INTELSAT as essentially the amount of satellite power-bandwidth combination required to carry on one-half of a two-way telephone conversation at a quality level prescribed and sanctioned by the International Telecommunication Union.) No output unit briefer than the unit-year, such as a unit-month, was considered in Chapter 2. Yet all of INTELSAT's so-called occasional usage consists of traffic usage purchased in blocs of time much briefer than a year. Television service, for example, is tariffed by the minute, and short-term units of utilization for cable restoration and other purposes are tariffed by the day. Evidently, there is a vertical (instantaneous) as well as a horizontal (through time) dimension to INTELSAT's output: at any given moment a certain power-bandwidth combination is being provided, and it continues to be provided for a given amount of time (the analogue in electricity generation is the kilowatt-hour). But there is a tacit linearity of an important and nontrivial sort in the very use of the expression "unit-year" or any other such two-dimensioned measurement of duration times instantaneous output. It implies, for example, that six units over a period of two months are indistinguishable from one unit over a period of a year, or, in an extreme example, from 525,600 units over a period of sixty seconds, indistinguishable being used in the sense of a measure of output in the various cost equations.

There were quite compelling reasons to abstract from output measurements shorter than one year in deriving the cost functions in Chapter 2. First, the bulk of INTELSAT traffic has historically consisted of units of utilization leased for one-year periods for a given one-year tariff (presently, however, the minimum full-time lease period is one month). Second, the data were not sufficiently copious or minutely calibrated in time to allow, for example, a month-by-month reconstruction of costs for a wide variety of services. Even if this had been possible, random and systematic fluctuations would have required lagged regressions that might have been difficult to specify given the brief operating period and lack of cross-sectional data. Therefore, the year was chosen as the basic output and cost period, and costs and output were related to that time framework. It was found, of course, that given the measurement standard of one year, decreasing costs, or nonlinearities in cost and output, occurred: two units of utilization cost less than twice as much as one, three units cost less than half again as much as two, and so on. There is every reason to believe that this same decreasing cost relationship holds for time periods less than one year, although the cost equations in Chapter 2 do not permit such a conclusion in and of themselves. The qualitative assumption will be made that decreasing costs continue to apply for output periods briefer than one year, although no effort will be made to determine their extent. This assumption will be important in the following review of the INTELSAT tariff, many service categories of which are for periods of time much briefer than one year.

THE INTELSAT TARIFF: INDIVIDUAL SERVICES AND THEIR CHARGING BASIS

Single-Destination Television

The first television pricing decision came in 1965 a month before the Early Bird satellite entered commercial operation. The ICSC decided at that time to charge all "occasional" traffic, including television, on the basis of $3.75 per hour per unit.[4] The number of units needed for a one-way television channel was estimated to be 240 units for Early Bird or "Quality 1" transmission.[5] In actual practice, television usage was charged on the basis of the 480 units of maximum capacity in INTELSAT I (Early Bird), less the number of full-time paid units allocated on the satellite at the time of the telecast (this averaged about 340 units (480-140) during the last half of 1965). Five months later, while INTELSAT I was still the only satellite aloft, the hourly per-unit rate for television and other occasional services was reduced to $2.40, but the surrender and excess capacity charges persisted.[6]

The basis of the present television charging policy was established in July 1968, as the ICSC was considering policies for the INTELSAT III satellites to be launched the following year. Those satellites were to provide a higher, "Quality 2" grade of television service estimated to be equivalent to 432 units of utilization, exclusive of accompanying audio channels.[7] The ICSC, on the advice of INTELSAT's manager, prorated the $20,000 annual full-time charge per unit to 3.8¢ per minute and then rounded it to 4¢ per minute. Multiplied by 432 units, this came to $17.28 per minute for television, which when rounded to $17.50 and divided by two, amounted to $8.75 per minute at each end of the transmission. A ten-minute minimum was set, and accompanying audio channels were charged at 4¢ per minute.[8]

The rate of $8.75 per minute at each end is still the INTELSAT charge for single-destination television.[9] Two factors have intervened since the time of its initial establishment to reduce whatever link it originally had to notions of underlying capacity usage and costs. First, the annual full-time unit rate has fallen four times from the 1968 level of $20,000 to its present level of $9,000 (see Table 4.1). Second, INTELSAT IV television standards have generally been of the higher "Quality 3" type. The level of 720 units of utilization originally ascribed by the INTELSAT manager to that level of quality in 1968[10] now seems speculative, and there is no particular necessity of arriving at an exact figure. Suffice it to say that until recently, television transmissions via INTELSAT IV satellites have occupied an entire global beam transponder, which has a maximum capacity of about 932 units. Recently, however, INTELSAT IV television transmission have used only half a transponder with a consequent reduction in quality. In general, the level of quality is a function of available capacity and other operational factors that frustrate any attempt to equate a video channel with any particular equivalent number of voice channels or units of utilization.

Before examining the basis in cost, if any, of television charges established by INTELSAT (an exercise which will be repeated below for other short-term and specialized services), it should be noted, regarding the use of cost curves disaggregated for individual services generally, that while cost curves were derived by statistical and accounting methods in Chapter 2, they were obtained only under the assumption of a single output and for periods of one year or more. For this reason, the cost curves of Chapter 2 will not be used to draw conclusions regarding the basis in cost of the tariffed services discussed below, that is, services which are either specialized, shorter than one year, or both, thereby deviating from the assumptions on which the cost derivations of Chapter 2 were based. (An exception will be the treatment of NASA service.) No attempt, furthermore, will be made to derive cost curves explicitly disaggregated to reflect these services, since the data available would be even more limited and the assumptions even more open to conjecture than was the case in Chapter 2. Instead, qualitative judgments will be offered regarding the charging basis and philosophy of the different tariffs. Attention will be paid to the following main categories of charging philosophy: average (fully allocated) cost, incremental (marginal) cost, and value of service (price discrimination).

What, then, is the charging basis of the single-destination television rate? Now that the discussion focuses on costs attributable to one of a number of different services, the question of cost allocation arises, which was of no concern in Chapter 2 when a homogeneous output was assumed. As in other cases, judgments regarding allocation must necessarily be to some degree arbitrary, and no sure extraction of underlying average or marginal costs will be possible with joint production of multiple services of the nature experienced in INTELSAT.[11] In general, individual services will be assigned for costing purposes to subcomponents of individual satellites or to entire individual satellites themselves.

To begin grappling with this problem in the case of single-destination television charging, some comments made by the Finance Subcommittee of the ICSC may be instructive.[12] In May 1970, that body attempted to ascertain whether the television charge, then as now at the rate of $8.75 per minute at each end, was covering "fully allocated" or only "incremental" costs. To determine fully allocated costs of television service, the subcommittee arbitrarily allocated a total of 960 units, that is, one 480-unit channel in the Atlantic and one in the Pacific, to that service. Then, in equally arbitrary fashion, it calculated total commercial space segment capacity available as one-third more than the amount of capacity needed to satisfy the discounted traffic data base projections for the (future) years 1971-75, inclusive. In doing so, the subcommittee compared the presumptive television capacity not with overall satellite capacity at its most efficient use but at its much less efficient capacity given the carrier configuration that would be decided upon if the projected traffic were indeed to materialize. Television usage was assumed to increase at a rate of 20 percent annually during the period under consideration. Thus, for example, in 1973, television was considered responsible for 960 out of 15,500 available commerical units, and the revenue requirement for television was therefore set at 960/15,500 of the total revenue requirement of $102 million, that is, $5.9 million. At the assumed level of television usage in 1973, that amounted to $32 per minute at each end. The figure declined from $52 per minute at each end in 1971 to $16 in 1975 on the fully allocated basis described above. Thus, on a fully allocated basis, INTELSAT television rates, it was concluded, in no way came close to paying for television service. This conclusion was influential in keeping the rate at $8.75 per minute while other rates came down more or less in proportion to the full-time usage rate.

In calculating incremental costs for television service, the subcommittee regarded television as "having no greater effect on space segment costs than bringing forward the date by which further capacity needs to be provided."[13] This comes close to capturing the economic sense in which marginal costs are opportunity costs. As a consequence of the tacit assumption of output inelasticity of costs, however, no marginal costs in the sense of costs variable with the level of output were calculated for inclusion in incremental costs. The subcommittee concluded that as a direct consequence of television service, new capacity would have to be provided in the Atlantic and Pacific regions six months earlier than would have been the case without such service.[14] The extra annual cost of this earlier provision was calculated at $600,000, which, assuming total television usage of 2,000 hours annually, would mean a revenue requirement of $5.00 per half-channel of television per hour during 1971-75. This is of course below the $8.75 charge then and now in force. At this point there appears to be little ground on which to take issue with the Finance Subcommittee's conclusion:

> Thus there would be scope for a reduction in TV charges if the incremental cost approach were considered equitable, but if the fully allocated cost approach were considered appropriate, TV may be considerably under-priced.[15]

As the TV charge has stayed at the same level from 1970 through 1975, however, this has meant that television has come closer to paying its average cost as total average costs for the system have fallen.

Multidestination Television

Any number of earth stations are technically capable of receiving a television broadcast originating from another earth station. Furthermore, the addition of one receiving earth station to that or those already receiving a telecast occasions no additional costs to INTELSAT with the exception of minor extra billing and setup costs. In its initial consideration of how multidestination TV broadcasts should be charged, the Finance Subcommittee outlined the two major charging possibilities: to charge each receiving station, regardless of the total number of such stations the full $8.75 per minute, that is, treat each as if it were the only receiving station; or to effect some reduction to users to reflect costs, such as dividing the total charge of $17.50 per minute equally among the transmitting and receiving stations. The subcommittee noted that the former alternative would contravene a normal INTELSAT practice of charging based on average cost recovery:

> [If each receive station were charged the full amount] INTELSAT would virtually be adopting a principle of charging according to the mode of exploitation of facilities by the hirer of space

> segment facilities and consequently the value to the ultimate consumer. This is not a principle normally applied by INTELSAT in fixing charges.[16]

"Value-of-service" pricing, of course, is a polite name for price discrimination, which is generally based on differences in the price elasticities of demand of different groups of customers. Here it is encountered for the first time in relation to INTELSAT. Price discrimination will be discussed in some detail in Chapter 5. Despite its realization of the discriminatory nature of the former alternative, the Finance Subcommittee recommended it due to uncertainties in future traffic and the extra administrative costs it felt might be incurred. The ICSC as a whole concurred in that recommendation, deciding on an interim basis

> that there would be one charge at the transmitting end while each receiving station would be charged an amount equal to that which would be applied if it were involved in only a single destination TV transmission.[17]

Finally, in response primarily to complaints of the Spanish[18] and Latin American administrations, the ICSC decided that receiving stations in multidestination broadcasts would only have to pay 50 percent of what each would pay if receiving a single-destination broadcast.[19] This policy is still in effect in the current INTELSAT tariff.[20] It should be noted that the present multidestination TV policy reflects a cost-based, nondiscriminatory price only in the case of two receiving stations, in which case the same revenue is recovered as would be in a single-destination broadcast. To recover the same revenue ($17.50 per minute) as in single-destination broadcasts for three or more receiving stations, the per-minute charge could alternatively be $\$17.50/(n + 1)$ for the sender and for each of the n receivers, or, if the sender were still required to pay $8.75 per minute (as is presently the case), the n receivers could each be charged $\$8.75/n$ per minute. Thus, for $n \geq 3$ receiving stations, the present INTELSAT tariff for multidestination television broadcasts still reflects a value-to-user philosophy. It bears noting that INTELSAT chose perhaps its most price-inelastic service for which to charge on a value-to-user basis.

Occasional Use Other Than for Television or Cable Circuit Restoration

INTELSAT users are able to lease voice-grade channels for various purposes for periods of 24 hours; Table 4.2 summarizes the charging pattern for occcasional use as it has appeared in successive INTELSAT tariff manuals.

The pattern is clear: the first day, or fraction thereof, of occasional service is charged at three times the equivalent annual rate; the day after

TABLE 4.2

INTELSAT Tariffs for Occasional Use Other Than for Television or Cable Restoration (Prices in current dollars)

Tariff Date	Utilization Charge: First Day[a]	Utilization Charge: Next Day	Utilization Charge: Next Eight Days	Utilization Charge: Additional Consecutive Days	Maximum Charge	Annual Full-Time Charge	Equivalent Daily Full-Time Charge
January 1966	167	167	83	67	—	20,000	55
January 1971	125	125	62.50	50	3,750[b]	15,000	41
January 1972	108	108	54	43	1,080[c]	13,000	36
January 1973	93	93	46.50	37	930[c]	11,160	31
January 1974	75	75	37.50	30	750[c]	9,000	25

[a]Or fraction thereof; "day" means 24-hour period.
[b]Over a 90-day period.
[c]Over a 30-day period.

Sources: INTELSAT Tariff Manual, January 1971, pp. 13-14, December 1971, pp. 13-14, December 1972, pp. 13-14, April 1974, pp. 12-13.

that costs the same; the third through tenth days each cost 1.5 times the equivalent full-time daily rate; and consecutive days thereafter cost about 1.2 times the equivalent full-time daily rate. The maximum charge provision, however, assures that the occasional user need pay no more than 20 consecutive days in a 30-day period (or 70 consecutive days in a 90-day period during 1971) after that he is considered a full-time user and is allowed to pay the lower full-time rate.*

It is evident that in its relation to costs, the occasional-use tariff is quite different from either the single-destination or multidestination TV tariff. The occasional tariff incorporates declining unit charges for increasing lengths of usage. Assuming that decreasing costs pervade even brief periods of usage, this means that the occasional-use tariff has a much closer relationship to underlying marginal and average costs than does the single-destination charge, which is based on a linear prorating of the annual charge into shorter periods, or the multidestination TV charge, which ignores underlying costs altogether for three or more receive stations in favor of a value-to-user charging philosophy.

Cable Circuit Restoration

Table 4.3 indicates the evolution of INTELSAT rates granted to cable owners in cases of unanticipated interruptions in the transoceanic telecommunications system.

As Table 4.3 shows, the daily rate for cable restoration circuits was equated to the pro-rata full-time charge until 1974, at which time it stayed at its old level when the annual full-time charge was reduced. One concludes, therefore, that the cable restoration rate is even more "promotional" or "concessionary" or far from recovering average costs than is the INTELSAT television tariff. This is because the TV tariff, although likewise derived pro-rata from the full-time charge at the $20,000 level, has, unlike the cable restoration rate, not fallen apace with the full-time charge to $11,160. If, therefore, the TV rate is nonetheless still well below average cost, the cable restoration rate must be even further below average cost. Again, cost allocation becomes arbitrary. Serving as it does unanticipated emergency traffic,

---

*This underlying rate structure was adopted at the Ninth Meeting of the ICSC in July 1965 and has been used ever since. It is based on CCITT charging principles, which provide that the first two days of any occasional telecommunications service should cost 10 percent of the (equivalent full-time) momthly rate; the following eight days should cost 5 percent of the monthly rate; and each day thereafter should cost 4 percent of the monthly rate: see ICSC-57-20, February 2, 1972, p. 2. This useful document summarizes the history, philosophy and then current status of charges for all three occasional use services: television, cable restoration, and other occasional uses.

TABLE 4.3

INTELSAT Tariffs for Cable Restoration
($)

Tariff Date	Daily Charge	Annual Full-Time Charge	Equivalent Daily Full-Time Charge
January 1966	58	20,000	55
January 1971	43.50	15,000	41
January 1972	36	13,000	36
January 1973	31	11,160	31
January 1974	31	9,000	25

Sources: INTELSAT Tariff Manual, January 1971, December 1971, December 1972, and April 1974.

an a priori cost-causational allocation of capacity to cable restoration is something of a logical contradiction. In any event, the low level of that rate no doubt prompted the INTELSAT Board of Governors to decide not to reduce it again in 1974.[21]

Just because a tariff is below average cost does not mean that it is a marginal cost tariff. The cable restoration rate, for example, hardly appears to fall into that category. Despite the presumption of decreasing costs, the per-day charge does not decline with time as is the case with the regular occasional-use charge. A charge such as this would appear to offer the worst of both worlds: being less than the average cost it yields an operating deficit but being nonmarginal it does not compensate by offering any marginal optimality properties. To the extent it is equal to or greater than incremental cost, of course, it provides net revenues to the system which can be used to pay off general costs.

Finally, it is notable that the cable restoration rate goes in the opposite direction of what has been called perfect price discrimination,[22] that is, the tailoring of charges to the demand elasticity of each consumer. One would suppose that INTELSAT occasional users whose cables had temporarily failed would gladly pay higher rates than other customers to have their service restored during repairs and thereby avoid having to forfeit substantial revenues. Yet the restoration rate is lower not higher, than the regular occasional-use rate. This may be because most INTELSAT members consist of national post and telecommunications entities or other operating organizations which are responsible for both cable and satellite communications; often there is no particular financial separation of accounts. Such bodies would naturally be under instructions from their governments and legislatures to provide for continuity of service in the international network by making emergency restoration between modes as easy and inexpensive as possible.

Double-Hop Circuits

Since only about a third of the earth's surface is visible to a synchronous satellite, and earth stations on opposite sides of the globe therefore cannot communicate via a single satellite, there has been a need for satellite double-hop arrangements. For example, India and Pakistan cannot communicate with the East or West Coast earth stations in the United States via a single link. Likewise, earth stations in South America cannot reach earth stations in Asia or Australia via a single satellite. There exists, however, the operationally feasible possibility of a two-satellite link via an intermediate earth station in order to join country pairs on opposite sides of the globe. In the first example, traffic could transmit via an earth station in Japan from an Indian Ocean to a Pacific Ocean antenna. In the case of South America, the link to Asia or Australia could be made via any European country which had both Atlantic and Indian antennas.

In practical terms, only a small portion of world traffic, potential or actual, is between such country pairs, and cable-satellite combinations are generally used for those links. INTELSAT, however, was obliged to improvise a number of satellite double-hop arrangements early on as the result of operational difficulties, particularly with television transmissions, and as a working policy no extra charge was levied for the second hop. Interest in a double-hop charge policy had developed far enough to prompt its consideration by the ICSC in late 1970. One stimulus of such interest was the fact that the exact positioning of a satellite is a matter of operational judgement by INTELSAT as a whole and is often beyond the influence of individual administrations. Mexico, for example, is at the extreme western edge of the Atlantic zone, and Iran is at the extreme eastern edge. Thus, a double charge for double-hop service under circumstances where a more favorable satellite positioning might have made possible a single-hop link appears unjust and arbitrary from the point of view of the adversely affected earth stations.

From the standpoint of charging philosophy, the double-hop issue provides an interesting case. Marginal costs to INTELSAT are essentially zero for the second hop (although the intermediate earth station incurs costs that are generally levied by that earth station's national administration on the sending and receiving entities outside INTELSAT's administrative and financial framework); yet average or fully allocated costs must by their very nature include an extra component to represent the additional capacity in the second satellite. In November 1970,

INTELSAT arrived at a compromise decision to grant a single-hop charge for double-hop circuits only "during any space segment outage which requires a double-hop transmission to restore service"[23] but not for regular traffic. The more general charging alternative met opposition, among other places, in the United States from the FCC and the Department of State, which were lukewarm toward such a marginal-cost charging policy for fear of the precedent it might set. Other countries were fearful that the favorable satellite rate might drain traffic off of existing cable links. The double-hop policy set in 1970 is still in effect today.[24]

Minimum Charging Periods, Reservation Charges, Cancellation Charges, Program Charges

A detailed inquiry into this relatively minor area of INTELSAT charging policy would not be particularly worthwhile. Most charges of this nature, however, reflect marginal cost pricing to the extent they are implemented. Occasional-use charges impose a minimum 24-hour period, and the TV tariff imposes a ten-minute minimum. From time to time there has been sentiment for imposition of reservation, cancellation, and program charges in connection with TV broadcasts, in order to recover administrative costs incurred regardless of program length. For example, the ICSC's Finance Subcommittee has estimated directly allocable administrative and handling costs of about $30 per TV program.[25] In addition, it was argued that TV users are responsible for "restrict[ed] earning capacity" during the 15-minutes-per-program average required for satellite lineup and preparation before the broadcast can begin whether or not other traffic is actually preempted during that period.[26] To recover both the marginal administrative costs and fully-allocated lineup costs, the Finance Subcommittee recommended a program charge of $150 at each channel end for every TV broadcast and elimination of a 10-minute minimum.[27] Symptomatic of INTELSAT's eclectic approach to charging policy, the program charge was justified with both incremental and fully-allocated-cost arguments. No action was taken on this or similar proposals. As noted previously, only the minimum-use times for TV, circuit restoration, and other occasional-use charges are presently in the INTELSAT tariff.

Bandwidth-Based Charging

Regarding INTELSAT charging based on bandwidth efficiencies, it should be noted, first, that in Chapter 2, the assumption was made that each satellite was a fixed plant in the short run, and no particular inquiry was made into the technology whereby individual satellites produced output. Since the interseries long-run curves showed marked economies of scale, it would not be surprising to find such economies or other efficiencies in the production patterns of individual satellites

TABLE 4.4

INTELSAT IV Carrier Sizes, Bandwidth Requirements, and Efficiencies

Global Beam				Spot Beam			
(1) Carrier Size in 4 kHz Channels	(2) Carrier Bandwidth (mHz)	(3) Bandwidth Per Channel (kHz)	(4) Percent of Efficiency of Bandwidth Usage	(5) Carrier Size in 4 kHz Channels	(6) Carrier Bandwidth (mHz)	(7) Bandwidth Per Channel (kHz)	(8) Percent of Efficiency of Bandwidth Usage
24	2.5	102.4	3.8	60	2.5	41.7	9.6
60	5.0	83.3	4.8	132	5.0	37.9	10.6
96	7.5	78.1	5.1	192	7.5	39.1	10.2
132	10.0	75.8	5.2	254	10.0	39.7	10.1
252	15.0	59.5	6.7	432	15.0	34.7	11.1
—	—	—	—	612	20.0	32.7	12.2
432	25.0	58.0	6.9	792	25.0	31.6	12.3
972*	35.0	37.0	10.8	1,872[a]	35.0	19.2	20.8

*Full transponder

Notes: kHz = kilohertz; mHz = megahertz; Column (3) = Col. (2) ÷ Col. (1); Column (4) = 4 kHz ÷ Col. (3); Column (7) = Col. (6) ÷ Col. (5); Column (8) = 4 kHz ÷ Col. (7).

Source: Various documents of the Technical Subcommittee of the Interim Satellite Communications Committee.

as well. A closer look discloses that marked nonlinearities of cost and production are in fact present in all satellite series because of the nonlinear way in which total bandwidth is related to usable bandwidth for a particular level of output as measured in units of utilization. Attention will be focused below exclusively on the INTELSAT IV series, although the same basic patterns exist for earlier series as well.

Each INTELSAT IV satellite consists of twelve so-called transponders, which are its basic communications modules or building blocks. Four transponders are in the "spot beam" mode, which means that the power they radiate back to earth is focused densely onto a relatively small area. Another four are in the "global beam" mode, meaning that their power is radiated over a broader area (essentially all of the earth's surface visible to the satellite) but at much lower power levels. The remaining four can be switched to either type of beam, depending on system requirements. The components of the transponders are called carriers, from which the individual 4 kHz telephone-quality channels are derived. Each carrier provides a given number of channels and requires a given amount of bandwidth; these figures differ in the global and spot beam transponders.

As Table 4.4 indicates, the bandwidth requirement per marketed voice channel declines as the carrier size from which the channel is abstracted becomes larger, though by no means in a uniform pattern. Most of the reduction in relative wasted bandwidth occurs in the last jump between the subtransponder and the transponder size. The reason for these increases in efficiency of bandwidth utilization that go along with increased carrier size derives from the guard bands of frequency which are required to keep one actively used 4 kHz channel from interfering with others. The larger the number of channels directed from one destination to another, the smaller the relative amount of bandwidth required to protect against such interferences.

Here an analogy suggests itself; it will be pursued at some length in order to help infer conclusions regarding charging based on bandwidth efficiencies. Suppose, then, that a chemical firm produces a particular caustic acid which is dangerous and must be packaged in bulky, expensive containers for shipment, storage, and eventual use. To approximate the satellite case, make the otherwise contrary-to-fact assumption that the acid itself has no cost to the seller independent of its container. Assume also that average costs of packaging are constant when packaging is measured in terms of, say, weight. Finally, assume, as seems reasonable, that as the amount of acid purchased and shipped to an individual customer increases, the amount (and thereby the cost, given constant costs) of packaging required per gallon of acid decreases. The customer does not care how much packaging he buys; the number of gallons of acid is the significant variable from his standpoint. Yet the producer, anxious to recover costs, sees that his costs are directly sensitive to the amount of packaging he sells (from which the acid is derived). In determining his cost structure, the producer sees constant average costs in terms of packaging material but decreasing average costs in terms of gallons of acid and accompanying packaging. What pricing structure would he wish to adopt?

Consider, first, the case in which he offers acid at a constant price per gallon, regardless of the amount purchased. Suppose he knows how much will be purchased in which denominations (or can make a guess in the presence of rather inelastic demand) and sets the constant unit price, under the threat of regulatory constraint, just high enough to recover his costs plus a desired return. His guess proves correct; the given number of gallons are sold at the price he sets. Some sales are big and some are small, of course, He has recovered his costs plus a return so he might be presumed to be satisfied. Some of the purchasers, however, are probably less than satisfied. The large purchasers have paid more than the average total cost of the product they are buying, and so, according to a generally accepted criterion set forth by Kahn,[28] are subsidizing the small purchasers, who can buy for below average total cost. The pricing scheme is, of course, discriminatory, since prices clearly diverge from underlying costs however measured.

Consider now a second case, in which the (monopolistic) producer establishes a tariff reflecting the cost of production plus a desired return for each size of sale measured in gallons. This would mean declining unit prices per gallon for larger purchases. Here, assuming constant costs of packaging material, the producer need not know the size distribution of purchases, since costs will be recovered in any event. If the demand structure is the same as before and has some price elasticity, there should be more sales to large purchasers and fewer sales to small purchasers than under the former price regime. Total revenue may be more or less than previously, depending on the price elasticities of different classes of buyers. On the demand side, large purchasers are happier and small purchasers are less happy than in the constant-price case.

This is in many ways a curious situation. The constant-price tariff alternative, for example, is price-discriminatory, but only by accident would it be the revenue-maximizing price discrimination pattern for the producer. If demand of larger purchasers is much less elastic, for example, than that of small purchasers, the producer may even maximize revenue by charging more per unit for large than for small purchases despite the fact that the costs went in the opposite direction. (Large purchasers may get around this difficulty by dividing their purchases up into several smaller components.) From the overall standpoint of allocative welfare, however, one point is clear: under the second charging regime, embodying cost-based, declining-unit prices, and assuming no externalities, the cost to society of producing each marginal unit is paid back by the person receiving the benefit of those units. An enlightened regulatory body would therefore impose the second rather than the first tariff described above.

Yet INTELSAT, which subsumes producers, purchasers, and regulators of satellite communications in one organization, markets full-time units of satellite utilization at a constant price, although its unit cost structure within individual satellites is decreasing (as Table 4.4 indicates). It was concluded in Chapter 2 that satellite capital costs between series were decreasing, and that variable (administrative) costs were decreasing for any amount of traffic distributed in any pattern. Thus, one more dimension of the pervasive decreasing cost structure of international satellite communications has been established by refining the conceptualization of the production process.

In terms of INTELSAT's tariff structure, the consequences are clear. Above and beyond any nonlinearities in cost among separate satellites of the same or different series, such nonlinearities exist within individual satellites, at least in the INTELSAT IV series discussed here. Consequently, ceteris paribus, INTELSAT would allocate its resources most efficiently on an overall basis by charging according to a declining unit schedule. Such a schedule could be "smoothed" to eliminate peculiarities of carrier assignment due to technical and operational factors beyond the control of individual users.* A smoothing of the schedule, for example, might eliminate separate lower unit charges for the spot beam, since the aiming of the spot beam is at the discretion of INTELSAT's operational experts, who seek to maximize overall system coverage and efficiency. Ideally, the smoothed charging schedule would not contain a single per-unit price for each size of purchase but would instead be a stepwise declining rate with the first 24 (or 60) channels selling at a given unit price, the next 96 (or 132) at a lower unit price, and so forth. While this would have a desirable overall allocative effect, how-ever, Pareto optimality would not result. Small users, heretofore paying the same rate as large users, would have to pay more and would

*These factors are not inconsiderable and have been used to argue against bandwidth charging of any kind. (1) bandwidth is strictly speaking not the only relevant parameter, as some carriers yield more units from given bandwidth when power is increased; (2) the carrier configurations of individual satellites can and have been revised from time to time depending on overall system requirements; (3) earth station owners would have an incentive to squeeze as many units as possible from bandwidth purchases, thereby decreasing system quality; (4) special modes of satellite modulation such as PCM/PSK, with single-channel-per-carrier operation, are not as subject to variations in relative bandwidth efficiency; and (5) in general the decentralized decisions of earth station owners might conflict with overall system optimization requirements, particularly when earth station expenses as well as space segment (INTELSAT) expenses are considered.

therefore be worse off. The political implications of raising the rates to mostly underveloped countries, many of whom joined INTELSAT only after committing several million dollars of scarce foreign exchange to the construction of an earth station and the purchase of investment shares in INTELSAT, are considerable.

Charges for Whole and Half-Transponders

In any event, INTELSAT has decided against a bandwidth-based pricing structure below the transponder or half-transponder level. It is to these latter tariffs that attention will now be directed. In the analogy of the acid manufacturer used above, it was seen that the producer himself had no particular incentive to price according to costs (apart from using MC = MR as a monopolist's guide to profit maximization), but that an impartial regulatory body with overall welfare in mind would impose a cost-based tariff. Yet the analogy is not representative of INTELSAT to the extent that it does not consider situations of excess capacity, which occur in INTELSAT with some regularity. Given excess capacity, a producer has his own incentive for selling at prices as low as marginal costs, aside from any overall welfare benefits of marginal cost pricing.

The question of marketing excess capacity is raised at this point because the only instances of INTELSAT marketing capacity on what might appear to be a bandwidth basis, namely a number of recent transponder and half-transponder leases, can also be interpreted as sales from excess capacity at below average cost. The details of the individual sales will be presented below after a discussion of INTELSAT's views toward whole-transponder sales of capacity. It should be noted, however, that the sale of any services in terms of transponder units or subunits such as half-transponder is bandwidth-based pricing rather than unit-based pricing. The extent to which the actual prices charged for the transponders or their submultiples reflect underlying costs (however conceived) is a separate question.

The Finance Subcommittee first touched on the question of marketing spare capacity in March 1971:

> Circumstances [might] arise . . . in which an incremental cost approach might be thought justified in the interests of earning revenue from otherwise idle capacity. But it would be very important to avoid creating serious anomalies by comparison with other INTELSAT tariff offerings. If, for example, transponders were made available to meet special requests it would be unreasonable to deny Signatories already using the space segment the opportunity to make similar application, and if the rate were a favorable one it would enable Signatories with large voice circuit requirements to economize, at the expense of INTELSAT revenues, in their space segment charges.[29]

In September 1971, the Finance Subcommittee arrived at figures of 360 units and 180 units as the recommended full-time charges for a whole and half transponder, respectively, for television broadcasts via INTELSAT IV satellites.[30] Those rates were based on quite arbitrary assumptions of average or fully-allocated pricing, according to which "the average revenue-producing capacity of an INTELSAT IV global beam transponder is 360 units—this is 80 percent of the 450-unit capacity which has been estimated by the ICSC/T [Technical Subcommittee]."[31] The 450-unit capacity determined by the ICSC's Technical Subcommittee is in conflict with the maximum global-beam transponder capacity of 972 units shown in Table 4.4. The lower capacity, therefore, must have been based on assumptions of less than maximally efficient traffic patterns broken down into smaller, less efficient carriers.

In any event, it is of interest that the fully-allocated-cost approach whereby the Finance Subcommittee arrived at the figure of 360 units for a whole transponder—whatever its merits—also comes close to reflecting underlying bandwidth efficiencies. As Table 4.4 indicates, a global beam transponder produces a maximum of 972 voice channels or units of utilization when used as an entire carrier, and requires only 0.361 (37.0 ÷ 102.4) as much bandwidth per channel as the smallest (24-channel) carrier. If usage and pricing were based on the smaller, inefficient carriers, the equivalent bandwidth-based charge should be 351 channels (972 × 0.361). Thus, the charge of 360 units suggested by the Finance Subcommittee comes close to approximating the underlying bandwidth efficiency cost savings of the full transponder when it is not broken down into smaller carriers. The half-transponder charge of 180 units is more generous, since the usage of half a transponder is more inefficient on a bandwidth per-channel basis than is that of a whole transponder.

This 360-180 basis of transponder charging has been applied in at least three decisions of the ICSC and the INTELSAT Board of Governors regarding transponder leasing to individual countries or groups of countries. In April 1972, in its first such decision, the ICSC applied the 360-unit charge to an allotment request by Brazil of a global beam Atlantic transponder on the major-path INTELSAT IV satellite for domestic use.[32] Later, in June of that year, the ICSC leased to Spain and Mexico a full-time television channel for a charge of 180 units for a minimum period of one year, the assumption being that the quality of television involved would occupy one half of an INTELSAT IV transponder.[33] Finally, in October 1973, the Board of Governors decided to lease a full-time global beam transponder to Comsat for use between Hawaii and the U.S. mainland. The charge was set at 360 units of utilization for a 21-month period.[34]

These initial decisions, as noted, were based on the Finance Subcommittee's concept of fully allocated pricing, which incidentally also approximates bandwidth pricing in this case, since the units of sale (transponders and their submultiples) are defined in terms of bandwidth rather than derivable voice channels.

More recent decisions in this area, by contrast, have approached incremental cost pricing or even price discrimination based on relative price elasticities of demand for different categories of users. The pioneer decision in this latter area involved the lease of a full-time global beam transponder to Algeria out of spare capacity, in contrast to earlier leases out of major-path satellites. In September 1973, Algeria requested a long-term global beam lease out of spare capacity at a rate of $1 million per year for domestic-only use on an interruptible basis, to begin in early 1975. This rate, of course, was well below that for the fully allocated transponder rate of 360 units, which in 1974 would have been $3,240,000 per year (360 × $9,000). As justification for the lower rate, Algeria noted that "the spare capacity used would otherwise generate no revenue," and that "the proposed transponder charge will make a very substantial contribution to INTELSAT's annual revenue requirement throughout the lease period to the financial benefit of all users of the system."[35] One could argue, on the one hand, that the service requested was of a lower quality than that purchased in previous transponder leases, in that it was to be interruptible without compensation and would be carried on spare capacity rather than on a major-path satellite. On the other hand, however, the rationale of the proposed charge was clearly based upon considerations of marketing spare capacity through price discrimination. Assuming, indeed, that Algeria's demand for domestic services is more elastic than demand generally for INTELSAT circuits, the arguments advanced by Algeria and subsequently adopted by the Finance Subcommittee and the Board of Governors (see below) become classical statements of the arguments in favor of price discrimination as a means of increasing revenues in the presence of spare capacity and decreasing costs.

The Finance Subcommittee, considering the Algerian request, stated that "most" of its representatives

> agreed that it would be in the interest of INTELSAT to establish . . . a new type of space segment utilization for domestic services, using spare capacity at a reduced charge. It seemed reasonable . . . to expect that by this means traffic could be attracted to, or retained by, the INTELSAT system on a scale which would improve the financial position of INTELSAT as a whole and effect a reduction in the space segment cost for each user in the whole system.[36]

The Board of Governors agreed, in October 1973, to grant Algeria the requested global transponder for domestic use on a preemptible basis, along with a number of other explicit restrictions not applicable to regular INTELSAT usage, and in particular not applicable to earlier major-path transponder leases. The price, as requested, was set at $1 million per year, to extend for a period of 5 years beginning January 1, 1975.[37] Assuming mid-range estimates by the Finance Subcommittee regarding the full-time unit charge during those years,[38] the $1 million annual charge would yield about half as much revenue as a 360-unit transponder charge during 1975-79.

Along the lines of the lease to Algeria, the Board of Governors approved long-term leases to Norway and Malaysia in June 1974. The lease to Norway was for a half-transponder out of spare capacity for domestic purposes, at a price of $500, 000 per year for five years. Malaysia was granted a full transponder lease for five years at a price of $1 million per year.[39] Other countries, including Indonesia, the Philippines, Iran, Zaire, and Chile, have expressed interest in similar arrangements,[40] although the option of a separately-owned domestic system appeals to larger countries.

Multidestination Voice Frequency Telegraph (VFT) Bearers

This long title refers to a minor component of INTELSAT operation and revenue which is nevertheless of interest from a charging standpoint. The service is analogous to multidestination television in that the several telegraph channels abstracted from a voice channel can theoretically be received by any number of earth stations with no increase in marginal or average costs to INTELSAT aside from billing and other minor administrative expenses. Perhaps due to the small effect on its revenues, INTELSAT reached a decision in this matter opposite to its policy for miltidestination television; it decided, in August 1969, to charge the transmitting station as usual for the voice circuit according to the occasional rate, but to levy no charge on the receiving earth stations, regardless of their number.[41] This tariff, which is still in effect, is added to the small catalogue of INTELSAT tariffs which are truly of a marginal-cost character.

Program Channels and Data Transmission on an Hourly Basis

By mid-1972 enough requests were being received for allotments of units over quite short time periods, measured in hours rather than days, to prompt ICSC consideration of appropriate new tariffs. Up until that time a one-day occasional allotment was the shortest time period tariffed except for television. The ICSC decided in June 1972, to establish hourly tariffs for program (radio broadcast), data transmission, and other one-way uses (see Table 4. 5).

The Finance Subcommittee, which recommended these rates,[42] applied an extreme average cost criterion to arrive at the figures. The then prevailing $13, 000 annual unit charge was averaged to $40 per hour assuming usage of only one hour per day and no usage on

TABLE 4.5

Hourly Tariff, Program, and Data Transmission Charges
($)

	Capacity		
	8 kHz (2 units)	12 kHz (3 units)	16 kHz (4 units)
Single Destination			
Hourly Charge, Each End	30	45	60
Per Occasion Charge, Each End	15	15	15
Multidestination			
Hourly Charge			
Transmit End	30	45	60
Each Receive End	15	22.50	30
Per Occasion Charge, All Stations	15	15	15

Sources: ICSC-59-3, June 1972; INTELSAT Tariff Manual, 1974, p. 22.

weekends.[43] It was lowered to $30 per hour with the proviso of a $15 fixed charge.[44] It was urged not to divert traffic from the daily occasional usage by hourly rates which would be found more attractive.[45] It is notable that the rate levels in Table 4.5, nominally derived from the full-time rate in 1972, have not been lowered since that time despite two subsequent reductions in the full-time rate. Of primary interest in this tariff offering is the multidestination tariff, which like the multidestination TV tariff after which it is patterned, becomes price-discriminatory for three or more receive stations.

One-Way Service

On the recommendation of the Finance Subcommittee,[46] the ICSC adopted, in October 1972, a tariff for one-way FDM/FM transmissions of 8, 12, and 16 kHz. These were charged at the rates of 1, 1 1/2, and 2 units at each end, respectively, corresponding to the 4 kHz bandwidth of the full-time unit. Of interest here is the multidestination rate for such channels: the transmit station was to pay the same unit rates but each receiving station was to pay only half of the above amounts.[47] Once more, the multidestination rate, patterned after the multidestination TV tariff, is found to be price-discriminatory for three or more receive stations. Occasional daily rates were established along the lines of two-way occasional rates.[48]

Regarding charging for special types of satellite modulation and access, for present purposes, as before, the main point of interest is how their tariffing relates to underlying costs.*

Demand Assignment

Very roughly, demand assignment is a method of accessing a satellite by means of which two earth stations establish a circuit from a preexisting pool of circuits only when they wish to communicate, rather than leasing the needed capacity on a full-time basis. It is not unlike the "party line" in telephone service. Demand assignment is useful for weak traffic links which would not justify full-time allotments, and to large users at peak load times who find that all their full-time units are occupied. A system of demand assignment called SPADE was designed by Comsat Laboratories and is not in operation in the INTELSAT system.

Possible INTELSAT tariff policies with respect to demand assignment were discussed extensively by the ICSC and its Finance Subcommittee well before SPADE became available even on a trial basis. The two main possibilities were a charge per minute of use and a full-time access channel charge, with some variants including hybrids of the two.[49] In discussing the merits of each, an average cost approach was taken. The Finance Subcommittee, for example, recommended a provisional per-minute charge of between 15¢ and 30¢ at each end for SPADE access on the basis that this charge would recover an average of $10,000 per year at the time the full-time space segment charge was also $10,000 per year.[50] The access channel charge had the

---

*It should be noted that increases in satellite capacity by means of special access and modulation techniques constitute a form of technological progress usually called efficiency, and can be conceived of as an increase in the value of k in the production relation

$$q = kf(x_1, x_2, \ldots, x_n)$$

where q is output, the x's are the factors of production, and f is the production function proper. In other words, one makes better use of existing equipment rather than deriving economies from increases in factor inputs. It is noteworthy, too, that the expenses involved in implementing new modulation and access techniques fall to a great extent on the earth station owners. Although these costs are not borne by INTELSAT, their often significant effect on individual administrations has had a substantial influence on many of INTELSAT's operating and pricing decisions.

disadvantage of defeating, to a large extent, the purpose of demand assignment by requiring an outlay based on reserved capacity rather than carriage of actual traffic. Yet it did have the virtue of relating price to the cost of earth station equipment, which varied directly with the number of SPADE channels accessible.

In practice, both of these tariffs have been used. At first, the ICSC decided, in April 1970, to charge for SPADE usage on an access channel basis.[51] Some four years later, a charge of 15¢ per minute of satellite circuit holding time at each end was instituted.[52] The latter charge is presently in effect on an experimental basis.[53]

Preassigned single-channel, per-carrier PCM/PSK circuits refer to a satellite modulation technique whereby digitalized data are transmitted using only a single channel per carrier, thereby avoiding some of the bandwidth inefficiencies discussed above. SPADE, in fact, is a special adaptation of this technique.

As the offering of this service became possible via INTELSAT IV satellites, INTELSAT's charging approach to it began to develop. As in the case of demand assignment charging, there was some fear in May 1972 on the part of the Finance Subcommittee that cost-based charging for a new service might erode revenues from older, more established services:

> [Some subcommittee members] were concerned to avoid a situation in which a new INTELSAT charge, whilst perhaps in various respects entirely logical and cost-oriented, might lead to an erosion of present and prospective revenues from established tariffs. This erosion might prove to be of small dimensions, and in any event would have to be balanced against a more efficient use of space segment capacity.[54]

In an unusual move, the ICSC took action immediately thereafter despite the fact that the Finance Subcommittee, on which it had routinely relied for charging suggestions, had by that time reached no decision in this area. The ICSC decided on an average cost-based charge for two-way, point-to-point PCM/PSK data, levying the charge at one unit of utilization at each end provided the capacity used was no more than the 4 kHz in a standard FDM/FM circuit on which the unit charge was based.[55]

Later, requests were made for one-way PCM/PSK service tariffs, both for single and multiple destinations.[56] Comsat[57] and the Finance Subcommittee[58] both recommended tariffs similar to other service offerings. The ICSC decided along the lines of these recommendations to charge one-half unit at each end of the single-destination, one-way PCM/PSK transmission. For a multidestination transmission, one-half unit was levied at the transmit end and one-fourth of a unit at each receive end, as for multidestination television.[59]

Finally, it is of interest that the above tariffs were preceded by one for PCM/PSK data channels which used 24 times as much capacity as the single channel tariffed at one unit of utilization. In September 1971, the data channel charge was set at 24 units;[60] it has remained at that level since then. That decision was based on average cost recovery but certainly gave no discount for the efficiencies of larger blocks of usage, either within individual satellites or generally in the sense of the cost curves derived in Chapter 2 of this study.

Rate Adjustment Factors for Nonstandard Earth Stations

As was noted in Chapter 1, the payload weight constraints of rocket launchers used by INTELSAT have required that a large part of the communications burden be carried by earth stations, in terms of size, cost, and power. The relevant technical parameter for antennas is the ratio of gain to time; this ratio is 40.7 for the so-called standard earth stations which predominate in the INTELSAT system. These stations have antennas that are 80 to 100 feet across and cost several million dollars. The satellite unit of utilization, the basic building block of INTELSAT's tariff structure, is crucially based on assumptions regarding the gain/time ratio of antennas accessing the satellites.

For a variety of reasons, mostly related to cost and flexibility of operation, antennas with gain/time ratios lower than 40.7 have from time to time been granted access to the INTELSAT space segment. Physically, these antennas have smaller diameters, ranging from 12 feet upwards, and some are portable to at least some degree. For example, an unmanned nonstandard earth station has operated in Antarctica; former President Nixon's visit to China in 1973 was broadcast live via a nonstandard station provided to the Chinese government by an American firm; and experimental stations have been located on ships at sea and at Comsat headquarters in Washington.

To relate the demands on the space segment made by smaller antennas to those made by standard earth stations, a so-called penalty factor, or rate adjustment factor, has been used. It is essentially a ratio of the amount of power and/or bandwidth a nonstandard station requires to derive a given number of channels to what a standard station would require. The technical picture is more or less complicated depending on whether nonstandard stations do or do not work within the existing INTELSAT system as opposed to forming a separate network without interconnection.

Table 4.6 indicates theoretical rate adjustment factors for two nonstandard earth station sizes, each for global and spot beam operation and each for two types of satellite modulation.

For a given earth station G/T, therefore, the rate adjustment factor is by no means constant but instead varies both with the type of modulation and the beam configuration. These variations are particularly

TABLE 4.6

Nonstandard Earth Station Rate Adjustment Factors

Modulation	Beam	Rate Adjustment Factor for Antenna of G/T Shown: 40.7*	31.7	25.7
FDM/FM	Global	1	2.6	12.8
	Spot	1	1.6	2.6
SPADE	Global	1	2.3	8.4
	Spot	1	1.5	3.0

*Standard earth station.
Source: ICSC/F-23-3, March 8, 1971, p. 14.

wide in smaller earth stations. The Finance Subcommittee summarized the problems of charging philosophy posed by these rations:

> If charges were set up in these same ratios, the result would be that equivalent channels through non-standard stations would cost more through a global beam transponder than through a spot beam transponder. This would be discriminatory by favoring Signatories using a spot beam transponder, and would conflict with the policy of a common charge per unit in the regular system.
>
> On the other hand, if the global beam factor were used, it would mean that capacity leased for special services in a global beam transponder would be charged at approximately what it would cost if leased in the regular system, but if that happened to be in a spot beam transponder, the charge would produce more revenue for INTELSAT (particularly in the case of the 25.7 G/T station) than the same amount of spot beam capacity in regular INTELSAT service. The charge per channel would, however, be the same whether provided in the spot or global beam.[61]

Certainly, the "discriminatory" pricing alluded to above would be based on underlying cost realities and therefore not discriminatory in the economic sense. Considerations of equity, however, would militate against differentiating charges according to criteria over which the customer has no control, whether or not those criteria reflect underlying costs in the most minutely disaggregated sense. Once again, INTELSAT resolved the cost philosophy conflicts by using a grossly averaged cost as a tariff guideline: the adjustment factors of 2.5 and 10.0 were selected as guidelines for the 31.7 and 25.7 G/T earth stations, respectively.[62] These rate adjustment factors survive in the present tariff as "order-of-magnitude" suggestions with the proviso that the actual factors be set by the Board of Governors on a case-by-case basis.[63]

Free Use of the Space Segment

The extreme example of price discrimination-cum-internal subsidization is any provision for free use of a given facility when marginal costs are positive. Generally, INTELSAT has been quite strict in allowing free use of its space segment. The only categories of usage explicitly granted for no charge are all of a technical nature, involving tests and lineups of earth stations entering the INTELSAT system or expanding their operations therein.[64] All such usage is a necessary prelude to the subsequent establishment of paid traffic in the INTELSAT system.

There has always been some sympathy in INTELSAT for free usage of capacity for special events promoting awareness of international telecommunications, but that view has not generally prevailed. For example, early in 1969 the European Broadcast Union requested free use of the INTELSAT space segment for a "World Television Weekend" in 1970 but was eventually turned down by the ICSC.[65] This was due to fears of the precedent that might be created by such free usage, as well as to resentment that all earth station owners did not seem ready to waive their fees, which were quite large relative to the INTELSAT space segment to be waived. In contrast, however, a Swiss request[66] for free usage for a transmission from Switzerland to Canada commemorating the centennial of the International Telecommunication Union (ITU) was granted,[67] on the basis that the program was "brief, non-commercial, communications-oriented," and that all earth stations involved had also agreed to waive their fees. In addition, dedication programs of new earth stations are normally carried without charge, often with imaginative and colorful international broadcasts, but these are considered preoperational, and free usage is granted under the usual rubric of testing.

NASA Service

The evolution of the INTELSAT system was greatly affected by an agreement reached in July 1965, between INTELSAT and four of its members (the United States, Great Britain, Australia, and Spain) acting on behalf of NASA. The so-called NASA or NASCOM service which resulted from the agreement was used primarily for communications support of the Apollo lunar exploration program and, among other functions, provided the capacity whereby man's first step on the moon was telecast live back to earth. Much of the early traffic of the INTELSAT system consisted of this NASA traffic, and it is not an exaggeration to say that it was NASA requirements which exercised the predominant influence on the design, scheduling, and early operation of the INTELSAT II series.* Table 4.7 summarizes NASA revenue and traffic during the period 1967-71.

Originally, INTELSAT negotiated an agreement whereby the four members retailing the service to NASA would pay $25 million to INTELSAT during the three-year period beginning September 30, 1966.[68] This amount was later renegotiated to $21,208,000 since INTELSAT was unable to produce its services on time: the first INTELSAT II was not operationally available until February 1967. A number of INTELSAT members, notably France and Switzerland, charged that the rate was preferential to the four abovementioned INTELSAT members to which it was granted. As time for the renewal of the service approached in 1969, there was opposition to the idea of continuing on those terms a service which allowed special usage of degraded 3 kHz circuits, time-sharing of circuits, and use of the NASA capacity for commercial requirements by the four countries in the case of lack of spare capacity elsewhere in INTELSAT for television or cable restoration. Those INTELSAT members favoring the NASA arrangement argued that the traffic had helped to put INTELSAT on its feet and to provide stability for long-range planning. As Table 4.7 indicates, NASA traffic provided 39 percent of total INTELSAT revenue and 37 percent of total traffic in its initial year.

The Finance Subcommittee, asked to undertake a study of charging criteria for the proposed NASA service in 1965, arrived at a number of charging figures based on cost averaging. Specifically, the NASA charge was calculated as the residual of total system expenses after projected commercial revenues had been deducted.[69] The $25 million figure (mentioned above) seems to have been a very rough approximation to the subcommittee's recommendations. After the initial three-year period expired in September 1969, NASA traffic gradually fell to a trickle and was charged on a per-unit basis after corrections were made for degraded circuits and nonstandard earth stations.

Since NASA usage was so substantial in the early years of INTELSAT, perhaps the use of the aggregate short-run and long-run cost curves derived in Chapter 2 will help resolve the question of whether the four countries providing the NASA service received concessionary rates (in some sense of the word) from INTELSAT. This claim, of course, has its political as well as economic implications. NASA also provided launch services to INTELSAT and was the only entity technologically capable of doing so; in addition, there were the usual conflict-of-interest

*For example, the United States proposed that the four NASA service carriers be responsible for 75 percent of INTELSAT revenue requirements through August 1, 1968. See ICSC-11-3, August 1965, para. 37, pp. 12-13.

TABLE 4.7

NASA Revenue and Traffic in INTELSAT
(Revenue in $000)

Year	NASA Revenue 1965 Prices*	NASA Revenue Current Prices	Total Revenue at Current Prices	Percent of Total Revenue from NASA	NASA Traffic (units)	Total Traffic (units)	Percent of Total Traffic from NASA
1967	6,063	6,562	16,736	39.2	300	805	37.3
1968	7,480	8,348	28,882	28.9	393	1,420	27.7
1969	6,264	7,233	44,672	16.2	342	2,197	15.6
1970	3,040	3,672	75,375	4.9	185	3,698	5.0
1971	1,445	1,797	78,251	2.3	132	5,117	2.6
Totals	24,292	27,612	243,916	11.3	1,352	13,237	10.2

*Discounted by composite index.

Sources: INTELSAT Financial Reports (annual); INTELSAT System Status Reports (monthly and bimonthly).

charges regarding Comsat's dual role as INTELSAT's manager and as U.S. representative in INTELSAT.

Table 4.8 indicates the difficulties involved in assigning costs to particular services once overall cost curves have been estimated. Four individual methods of cost assignment to NASA service are suggested. In Method 1, NASA traffic is assumed responsible for the "first" or leftmost part of the aggregate total cost curve, and all other traffic is assumed responsible for the "rest"; in terms of equation 2.8, the short-run cost allocable to NASA in 1967 is AgSRTC(300) and that allocable to the remaining 505 unit-years of traffic is AgSRTC(805) - AgSRTC(300). Method 1 is therefore an extremely severe assessment of NASA's responsibility in total system costs, in that it includes all fixed costs and the most expensive variable costs (those to which declining costs are least applicable). Method 2 errs in the opposite direction, assigning NASA traffic entirely to the "marginal" category; no fixed costs and only the last or least expensive variable costs are assigned. Method 3, roughly that used by INTELSAT, assigns NASA communications cost as that proportion of total costs which NASA traffic comprises of total traffic. Method 4, applicable only to the short run, is, like Method 3, a kind of compromise between the extremes posed by the first two: in it, all fixed costs of the INTELSAT II system are allocated to INTELSAT, on the argument that that series was developed mainly for NASA traffic; variable costs are assigned on a fully distributed basis as in Method 3.

As Table 4.8 indicates, as much as 90.7 percent or as little as 1.0 percent of total INTELSAT costs during 1967-71 can be attributed to NASA service, depending on the cost allocation method chosen. The arbitrariness and great variation of the four methods illustrated underscore the difficulty of assigning costs under a regime of joint costs and production along with substantial decreasing costs, as noted earlier.[70] In the absence of convincing arguments to the contrary, the fully-allocated approach of Method 3 seems as sensible a compromise as any.

THE INTELSAT PRICING STRUCTURE: RETROSPECT

As noted before, the average cost philosophy predominates in INTELSAT ratemaking, although traces of incremental and value-to-user pricing can be found (see Table 4.9). There has been some consistency of approach to charging as it has developed since INTELSAT began operation; perhaps the best overall statement of INTELSAT's charging philosophy is contained in observations of the Finance Subcommittee at its March 1971 meeting:

We recommend that the method to be adopted as the norm should consist of dividing total system costs, or total apportioned costs of the capacity providing the service or facility in question, by the number of units of that service or facility expected to be requested. For this purpose, "costs" would include the appropriate compensation for use of capital . . .

We consider that, generally, a rate so assessed ought not to be increased by "value to user" considerations, that is to say INTELSAT ought not to become involved in questions of the profitability to the allottee of particular commercial applications of the capacity assigned to him. Such an involvement . . . would overlook the fact that the relationship between INTELSAT charges and the tariff to the eventual customer is an indeterminate one. If, however, rates set below [average total costs] are applied for some services or facilities, in furtherance of some long-term objective, any deficiency in revenue must be made good somewhere and INTELSAT may then have to consider which services or facilities can best bear this burdern, e.g., by incorporating a "value to user" element in the charge for them.

Unless INTELSAT is to suffer actual financial loss in respect of particular services or facilities, the revenues yielded by a particular charge must at least cover the incremental cost of providing the service or facility in question. By incremental cost is meant the additional cost incurred by such provision, over and above the costs incurred if that facility were not provided at all. In the short-term, e.g., where there is spare capacity in a satellite, incremental costs may be virtually nil, and even in the longer term . . . incremental costs may be well below the fully allocated cost . . .

Circumstances may arise in which it would appear to be of financial advantage to INTELSAT to cover only the incremental costs, or something between that and fully allocated cost, if that were the best that could be secured, i.e., if a higher charge would mean loss of desirable business. However, we think that such a policy must be exercised with care since:

(a) in the long-term, revenues might be insufficient to cover the costs of keeping the requisite space segment capacity available.

(b) the concession might stimulate similar applications from other sources which could not so cheaply be met, or lead

TABLE 4.8

Costs of the NASA-INTELSAT Program, 1967-71, Under Various Methods of Calculation
($000 Discounted to 1965)

Year	Method 1[a]		Method 2[b]		Method 3[c]		Method 4[d]	
	Long-Run	Short-Run	Long-Run	Short-Run	Long-Run	Short-Run	Long-Run	Short-Run
1967	13,854	20,621	2,333	612	6,874	8,141	—	12,410
1968	14,982	27,368	1,950	489	6,018	8,059	—	18,475
1969	14,390	37,244	1,182	286	3,842	6,207	—	17,727
1970	12,041	31,352	424	99	1,436	1,780	—	5,501
1971	10,919	29,595	238	54	814	898	—	245
Totals	66,186	146,180	6,127	1,540	18,984	25,085	—	54,358
Percent of Total Short-Run or Long-Run Costs, 1967-71[e]	52.9	90.7	4.9	1.0	15.2	15.6	—	33.7

[a]NASA traffic comes "first"—rest is marginal.

[b]All other traffic comes "first"—NASA traffic is marginal.

[c]NASA costs fully allocated according to percentage of NASA traffic in total traffic.

[d]NASA costs = all INTELSAT II fixed costs + fully allocated variable costs. Method 4 applies only to short-run costs since variable costs are involved.

[e]Total short-run costs for 1967-71 were $161,229,000 in 1965 dollars; total long-run costs for 1967-71 were $125,100,000 in 1965 dollars.

Source: Compiled by the author.

TABLE 4.9

Summary of INTELSAT Tariff Structure

Service	Charges	Predominant Charging Policy: Average Cost	Incremental Cost	Value to User	Comments
Full-time unit lease	$9,000/year	X	—	—	Basis of most other charges; now revised annually
TV (single-destination)	$8.75/minute, each end 10-minute minimum	—	—	—	Below average cost but above incremental
TV (multidestination)	$8.75/minute to transmitter $3.88/minute to each receiver 10-minute minimum	—	—	X	For 3 or more receive stations
Occasional use	$75 each, first 2 days $37.50 each, next 8 days $30 each, succeeding days Maximum $750 in 30 days	X	—	—	Declining unit charges show some reflection of decreasing costs
Cable restoration	$31/day	—	—	—	Well below average cost, since prorated full-time charge
Double-hop	No extra charge for second hop, contingency only	—	X	—	ICSC refused to extend policy to noncontingency use
Minimum charging period	a	—	X	—	Few such policies in effect.
Bandwidth-based charging	b	X	—	—	Not necessarily marginal cost pricing but a more accurate perception of underlying costs of all types
Whole and half transponder Brazil U.S. Mainland-Hawaii Spain-Mexico	350 units, whole transponder 180 units, half transponder	X	—	—	Bandwidth-based charging; unit equivalences derived from cost averaging
Algeria Norway Malaysia	$1,000,000/year, whole transponder $500,000/year, half transponder	—	X	X	Spare capacity on interruptible basis for domestic use over long periods; can also be viewed as discriminatory pricing favoring higher-elasticity customers

Multidestination VFT bearers	Usual charge to transmitter	—	X	—	Goes counter to other multidestination pricing policies
Program channels					
Single-destination	c	X	—	—	For 3 or more receive stations
Multidestination	c	—	—	X	
One-way service					
Single-destination	See INTELSAT Tariff Manual	X	—	—	
Multidestination	See INTELSAT Tariff Manual	—	—	X	For 3 or more receive stations
Occasional	See INTELSAT Tariff Manual	X	—	—	Declining unit charges show some reflection of decreasing costs
SPADE	15¢ minute at each end during actual usage	X	—	—	Used to be on access channel basis
Preassigned single channel per carrier PCM/PSK	See INTELSAT Tariff Manual	X	—	X	Value to use for multidestination broadcast with over 3 receive stations
Nonstandard earth stations	Rate adjustment factors based on added power needed relative to standard stations	X	—	—	Exact values determined on case-by-case basis
Free use of space segment	See INTELSAT Tariff Manual	—	—	X	Limited in practice; obvious case of cross-subsidization and price discrimination
NASA service	$27,612,000 over 5 years for 1,352 unit-years (see Table 4.7)	X	—	—	Degraded circuits and time-sharing allowed

[a]See page 80.
[b]See Table 4.4.
[c]See Table 4.5.
Source: Compiled by the author.

to pressures from the users of similar facilities for their own, cost-oriented, charges to be reduced to a similar level.

In other circumstances it might seem appropriate to set a rate below the level indicated by the normal method of assessment, with the objective of thereby stimulating demand so that, in the longer run, use of the space segment would be increased on a scale leading to lower costs per unit of capacity, and thus benefit all Signatories. We think that whilst such measures might be appropriate in some circumstances it has to be recognized . . . that space segment charges affect customer demand only very indirectly. In particular, no attempt should be made to "compete" with other media by artificially manipulating the INTELSAT tariff structure to the detriment of particular users.

Considerations which can properly be taken into account in deciding whether to depart somewhat from a purely cost-based rate include the following:

(a) The relationship with other existing INTELSAT rates. Care must be taken to avoid anomalies which will distort the pattern of demand in a way injurious to INTELSAT as a whole.

(b) The range of choice of facilities open to Signatories. This range places some constraint on the extent to which Signatories can match the facilities paid for to those actually necessary to meet local requirements, and some limit must be placed on the extent to which the general interests of INTELSAT override the interests of individual Signatories.

(c) Some forms of charging, whilst soundly based on costs might produce unsatisfactory side effects from the operational viewpoint . . . .

(d) The tariff structure must be reasonably simple to administer.[71]

Most of the charging principles implicit in the INTELSAT tariff are contained somewhere in the above statement. Marginal-cost pricing is recommended on the grounds of attracting additional business, but the alleged optimality properties of charging exactly marginal prices are not discussed. There is some appreciation for the interrelationship between price and demand, along with the claim (true in general but probably with numerous exceptions for individual services, particularly at low traffic levels) that the INTELSAT charge has little effect on ultimate demand by individual consumers of telecommunications services. In addition, the dynamic aspects of charging low rates in order to induce higher demand and thereby scale economies of production are touched upon.

A very sketchy model will now be given of the revenue requirements approach used by INTELSAT. Future costs are assumed fixed at level c, and future demand is assumed fixed at level d. The unit price used as a charging benchmark is then set at c/d. Note that this approach assumes perfect inelasticity of demand to price and of cost to output. Both assumptions seem dubious, particularly the latter. These assumptions may well have been responsible for the apparent overinvestment in capacity that INTELSAT has practiced since it began operation. A different average-cost approach, one which incorporates non-zero elasticities of demand to price and of cost to output, is easy to formulate:

$$\begin{aligned} \text{demand} &= f(\text{price}) \\ \text{cost} &= g(\text{output}) \\ \text{output} \times \text{price} &= \text{revenue} \\ \text{revenue} &= \text{cost} \\ \text{output} &= \text{demand} \end{aligned} \tag{4.1}$$

where f is the demand function and g is the total cost function. The function g has already been estimated in equation 2.12 in its aggregate long-run form; presumably f should pose no difficulties of estimation either. Since equation (4.1) contains five expressions and five unknowns, no difficulty would appear in theory in refining the average cost approach to revenue requirements used by INTELSAT and thereby matching expenditures and facilities more closely to demand, with the obvious allocative benefits this would entail. Note that these benefits would be independent of those of any marginal-cost pricing scheme which might be attempted, the benefits and drawbacks of which have already been dealt with in Chapter 3. Equation 4.1, in fact, is merely a mathematical statement of the following familiar verbal rule: produce the output at which long-run average costs are equal to demand, and charge long-run average costs (LRAC) for that output. One difficulty is encountered if the LRAC and demand curves cross at more than one point. Since INTELSAT's LRAC curve has been found in Chapter 2 to be constantly downward-sloping, there would appear to be no objection to selecting the intersection at which long-run average costs are lowest (that is, at which output is highest).

NOTES

1. INTELSAT Operating Agreement, Art. 8(a)-(b), pp. 293-94.
2. INTELSAT, International Communications Satellite Committee ICSC/F-23-3, March 8, 1971, p. 3 (hereinafter cited as ICSC).
3. See INTELSAT Financial Report (1972) April 18, 1973, app. A, for a summary of financial results from 1964 through February 11, 1973, including costs, return to capital, etc.
4. ICSC-8-3, June 1965.
5. ICSC-33-46, July 9, 1968, p. 1.
6. ICSC-14-3, November 1965.
7. ICSC-33-46, July 9, 1968, p. 4.
8. ICSC-33-3, July 1968.
9. INTELSAT Tariff Manual, April 3, 1974, para. 3(b), p. 14 (hereafter cited as INTELSAT Tariff Manual (1974).
10. ICSC-33-46, July 9, 1968, p. 1.
11. For an early treatment predating INTELSAT operation, see Leland L. Johnson, "Joint Cost and Price Discrimination: The Case of Communications Satellites," in William G. Shepherd and Thomas G. Gies, eds., Utility Regulation: New Directions in Theory and Policy, (New York: Random House, 1966), reprinted from The Journal of Business 37, no. 1 (January 1964), pp. 32-46. A random-orbiting system and link-specific charging are assumed.
12. See ICSC/F-20-3, May 21, 1970, para. 37-51, pp. 16-21.
13. Ibid., para. 47, p. 20.
14. Ibid.
15. Ibid., para. 49, p. 20.
16. ICSC/F-15-3, May 3, 1968, para. 17, p. 8.
17. ICSC-32-3, May 22, 1968, para. 50(e), p. 14.
18. See, for example, the proposal of Spain in ICSC-37-34, January 27, 1969.
19. ICSC-49-3, November 1970, para. 90.
20. INTELSAT Tariff Manual (1974), para. 3(d)(ii), p. 15.
21. The Finance Subcommittee's recommendations in this direction are in ICSC/F-30-3, January 12, 1973, paras. 35-44, pp. 15-20, and INTELSAT, Board of Governors BG/F-2-3, October 1, 1973, paras. 11-16, pp. 5-9 (hereafter cited as BG).
22. Alfred E. Kahn, The Economics of Regulation, vol. 1, Principles (New York: John Wiley, 1970) p. 132.
23. ICSC-49-3, November 1970, para. 78.
24. INTELSAT Tariff Manual (1974), para. 7, p. 19.
25. ICSC/F-25-3, September 15, 1971, para. 27(a), p. 10.
26. Ibid., para. 27(b).
27. Ibid., paras. 27-28.
28. Kahn, op. cit., p. 143.

29. ICSC/F-23-3, March 8, 1971, para. 32, pp. 10-11.

30. ICSC/F-25-3, September 15, 1971, paras. 10-22.

31. Ibid., para. 16, 3-8.

32. ICSC, 58-3, April 1972, para. 48. A later agreement, revised at Brazil's request, is contained in BG-5-64, October 29, 1973.

33. See details in ICSC-59-47, June 1972. See also INTELSAT Tariff Manual (1974), para. 13(a), p. 25.

34. BG-5-3, October 1973, para. 101; see also INTELSAT Tariff Manual (1974), para. 13(b), p. 25.

35. BG/F-2-11, September 27, 1973.

36. BG/F-2-3, October 1, 1973, para. 35, p. 15.

37. BG/F-5-3, October 1973, para. 93; INTELSAT Tariff Manual (1974), para. 13(c), p. 25.

38. Attachment No. 2 to BG/F-2-3, October 1, 1973.

39. BG-9-3, May 31, 1974, paras. 57-60, pp. 32-35.

40. Attachment No. 4 to BG/F-4-3.

41. ICSC-42-3, August 19, 1969, para. 57(a), p. 32; INTELSAT Tariff Manual (1974), para. 5(b), p. 17.

42. ICSC/F-28-3, May 26, 1972, para. 36, p. 13.

43. Ibid., para. 28, p. 10.

44. Ibid., para. 29, p. 10.

45. Ibid., para. 27, p. 10.

46. ICSC/F-29-3, September 15, 1972, paras. 15-16, pp. 6-7.

47. ICSC-61-3, October 1972, paras. 53(a)-(d); INTELSAT Tariff Manual (1974), para. 11, p. 23.

48. INTELSAT Tariff Manual (1974), para. 11(c), p. 23.

49. For an example of such discussions see the Finance Subcommittee's record in ICSC/F-17-3, May 16, 1969, paras. 10-21, pp. 6-10.

50. BG/F-2-3, October 1, 1973, para. 28, p. 13.

51. ICSC-49-3, November 1970, para. 91(d).

52. BG-5-3, October 1973, para. 47(a).

53. INTELSAT Tariff Manual (1974), para. 6, p. 18.

54. ICSC/F-28-3, May 28, 1972, para. 44, p. 16.

55. ICSC-59-3, June 1972, paras. 93(a) and 95; INTELSAT Tariff Manual (1974), para. 9(a), p. 21.

56. See U.S. contribution, ICSC-59-42, June 1972.

57. ICSC-59-43, June 1972.

58. ICSC/F-29-3, September 15, 1972, paras. 17-25, pp. 7-10.

59. ICSC-61-3, October 1972, paras. 54-55; INTELSAT Tariff Manual (1974), para. 9(b)-(c), p. 21.

60. ICSC-55-3, September 1971, para. 33; INTELSAT Tariff Manual (1974), para. 8, p. 20.

61. ICSC/F-23-3, March 8, 1971, paras. 46-47, p. 15. Emphasis in original.

62. ICSC-55-3, September 1971, para. 54.

63. INTELSAT Tariff Manual (1974), para. 3(b), p. 4.

64. Ibid., para. 6(a), p. 6.

65. ICSC-43-3, October 8, 1969, paras. 104-05, p. 83.

66. ICSC-39-19, June 1969.

67. ICSC-39-3, June 17, 1969, paras. 28-29, p. 12.

68. ICSC-12-3, September 1965.

69. See ICSC/F-4-3, September 1965. The entire Fourth Meeting of the ICSC/F was devoted to this question.

70. See Leland L. Johnson, op. cit.

71. ICSC/F-23-3, March 8, 1970, paras. 10-17, pp. 3-5.

CHAPTER

5

NEW DIRECTIONS IN INTELSAT RATEMAKING

In this chapter, two areas will be elaborated in which INTELSAT might attempt ratemaking innovations. The first, regional pricing, implies a disaggregative refinement of the average cost of service approach. The second, price discrimination, is a departure from average cost pricing to provide a means whereby operating losses incurred by pricing below average cost in high-elasticity services can be recovered in revenues from higher prices for low-elasticity services (as indicated in Chapter 2).

REGIONAL PRICING

In Chapter 1 it was observed that, apart from the expensive and risky maneuver of drifting satellites from one ocean region to another, the three ocean regions comprise essentially separate and nonoverlapping markets. The aggregative cost approach used in Chapter 2 was abstracted from this compartmentalization and dealt with the INTELSAT space segment as a whole. Yet if the three regions were considered separately, it would be only by accident that all three had the same cost structure.[1] If this is true, cost-based pricing would dictate different prices in each region. This will now be illustrated.

The time frame chosen for regional dissagregation of costs and rates is the three-year period from January 1971 through December 1973. Successively, prime INTELSAT III carriage of traffic was replaced by launches of INTELSAT IV satellites in each ocean basin. But no INTELSAT III satellites had yet passed their five-year in-orbit service period by the end of that time, and all were on active standby or could be used in an emergency.

Table 5.1 indicates the number of satellites of each type in each region during those months of 1971-73 in which there were increases, as well as during the first and last months.

TABLE 5.1

Total Number of Satellites, by INTELSAT Series and Region, 1971-73

Year	Month	Atlantic III	Atlantic IV	Pacific III	Pacific IV	Indian III	Indian IV
1971	January	3	0	1	0	1	0
1971	April	3	1	1	0	1	0
1972	March	3	2	1	1	1	0
1972	August	3	2	1	1	1	1
1973	September	3	3	1	1	1	1
1973	December	3	3	1	1	1	1

Sources: Communications Satellite Corporation, Pocket Guide to the Global Satellite System, March 1974, pp. 12-14; INTELSAT System Status Reports (monthly and bimonthly).

As variable costs were not disaggregated by satellite series, they will not be disaggregated by region either. At any given moment their total amount, by equation 2.1, is

$$1,440,000(\mathrm{TRAF})^{0.22}$$

in 1965 dollars per year. These variable costs are assigned regionally each month according to the proportion of total INTELSAT traffic in each region. This is in lieu of a true statistical disaggregation.

Total fixed costs in the short run are given in equation 2.2 for individual satellites, and these will be linearly aggregated by region the same way they were aggregated in equation 2.8 over the entire system. In this fashion, total costs will be recovered each month. Average costs will be calculated as usual by dividing by traffic, in this case by the traffic level in the applicable region. Table 5.2 indicates annual traffic levels during the period under consideration in three-month intervals.

Table 5.3 indicates average costs (total costs divided by traffic) on an annualized basis in 1965 prices for each region. For comparison, average costs for the entire system are given, as well as the space segment charge in effect during the period indicated.

An accepted criterion of cross-subsidization of services is that a service pay either less than its marginal cost or more than its average total cost.[2] By that standard, Table 5.3 can be interpreted as indicating that the Pacific region customers provided subsidy to the Indian region users for 10 out of 12 months and to the Atlantic region users for 6 out of 12 months. Likewise, Atlantic users provided subsidy,

TABLE 5.2

Total Traffic in Unit-Years, by Region, 1971-73

Year	Month	Atlantic	Pacific	Indian
1971	January	2,975	1,330	350
1971	April	3,105	1,305	410
1971	July	3,230	1,525	515
1971	October	3,210	1,565	645
1972	January	3,640	1,690	655
1972	April	3,800	1,775	685
1972	July	3,790	2,035	745
1972	October	4,435	1,840	815
1973	January	4,900	1,915	820
1973	April	5,120	1,985	930
1973	July	5,250	2,135	1,055
1973	October	5,915	2,190	1,205

Source: INTELSAT System Status Reports (monthly and bimonthly).

in each case to Indian Ocean users, in 6 out of 12 months. In two out of 12 months, Indian Ocean users subsidized Atlantic users; the rest of the time they themselves received subsidy. All users paid the space segment charge, of course, indicated in Table 5.3, instead of the regional average cost shown there.

Several observations may be made at this point. First, the assignment of satellites to particular ocean regions, and the timing of launches, is to some extent dependent on system planning considerations above and beyond the traffic requirements of those regions, and to that extent an element of unfairness inheres in any regional charging method. Second, the INTELSAT permanent agreements, as well as the U.S. Communications Satellite Act and various U.N. and ITU pronouncements, are replete with exhortations regarding the global nature of satellite communications, and can be interpreted as indirect recommendations against any regional pricing scheme. The INTELSAT Agreement, furthermore, contains specific injunctions that INTELSAT capacity be available on a "non-discriminatory" basis.[3] From an economic standpoint, there is no price discrimination when underlying cost relationships are not departed from, but it is likely that the global approach to INTELSAT might dictate global costing without regional disaggregation. Even more specifically, the INTELSAT Agreement provides that

> [the] rates of space segment utilization charge for each type of utilization shall be the same for all applicants for space segment capacity for that type of utilization.[4]

TABLE 5. 3

Average Costs by Region
(1965 prices in dollars)

Year	Month	Atlantic	Pacific	Indian	All Regions	Space Segment Charge
1971	January	6, 166*	5, 102*	13, 835	6, 439	12, 060
1971	April	7, 930	5, 109*	12, 047	7, 516	12, 060
1971	July	7, 568	4, 520*	9, 855	6, 909	12, 060
1971	October	7, 565	4, 412*	8, 192	6, 729	12, 060
1972	January	6, 748	4, 085*	7, 963	6, 129	10, 127
1972	April	8, 104	7, 396*	7, 630*	7, 851	10, 127
1972	July	8, 063	6, 594*	7, 084*	7, 497	10, 127
1972	October	8, 418*	7, 044*	14, 107	8, 715	10, 127
1973	January	7, 675*	6, 744*	13, 949	8, 115	8, 448
1973	April	7, 350*	6, 501*	12, 406	7, 726	8, 448
1973	July	7, 151*	6, 087*	11, 041	7, 368	8, 448
1973	October	6, 396*	5, 873*	9, 730	6, 704	8, 448
1971	12 months	7, 322	4, 762*	10, 505	6, 898	12, 060
1972	12 months	7, 601*	6, 591*	10, 837	7, 677	10, 127
1973	12 months	6, 725*	6, 283*	11, 559	7, 196	8, 448

*Regional average cost less than global average cost.

<u>Source:</u> Table 5. 2, equation 2. 9, and author's calculations.

Here is a Pandora's box of interpretation. For example, is the Pacific region's type of utilization different from that of Indian or the Atlantic region? Economically, the services are not substitutes.

There are certainly political realities in any scheme of regional pricing or regional subsidization, but they are not easily discernible. The main users in the Pacific are the United States and Japan, but underdeveloped nations such as Thailand and the Philippines are also users. Likewise, the Atlantic region's users include less developed countries such as Mexico, Greece, Senegal, and Nigeria. The Indian region includes developed users such as Great Britain, Australia, and New Zealand. On balance, however, the Indian region probably contains the greatest proportion of traffic from the less developed world and would therefore be the logical recipient of subsidy were regional pricing to be eschewed in favor of global pricing. Table 5.3 indicates that this has in fact been the case.

Finally, the economist's job is not to make value judgments about whether subsidization should occur, or what patterns it should follow, but to point out the economic costs of distortion and misallocation which must be weighed against the presumed benefits of global pricing. Furthermore, he must point out that even if subsidy is desired, it can be accomplished more efficiently by lump-sum transfers which do not distort the relationship of prices to costs.

In summary, while a regional pricing system would produce some additional economic benefit by aligning prices more closely with underlying costs,* political realities and the global conception of INTELSAT's mission embodied in relevant agreements, national legislation, and past practice would appear to preclude regional pricing as a realistic alternative to global pricing.

PRICE DISCRIMINATION

Price discrimination based on value-to-user considerations is specifically prohibited in the present INTELSAT Agreement.[5] Nevertheless, price discrimination has a certain attractiveness, particularly in decreasing-cost industries, the economic advantages of which have often been found to outweigh any drawbacks of seeming inequity as among groups of users. Furthermore, despite the prohibition of

---

*Note that the entire discussion in this subsection has dealt with average costs, abstracted from any marginal pricing questions. In addition, only the single service offering of full-time unit leasing has been considered. Finally, a number of substantial though not insurmountable operational and technical difficulties have not been mentioned.

discrimination in the agreement, INTELSAT has been found to have practiced forms of discrimination from time to time and the Finance Subcommittee has argued in its favor in certain situations, particularly those of idle capacity which could be sold at low rates to high-elasticity consumers.[6]

Regulatory agencies in the United States have permitted pervasive patterns of discriminatory pricing to emerge. Perhaps the most common examples are railroad freight rates which vary according to type of goods shipped, and differential electricity rates for domestic and industrial users, to the extent the latter do not reflect underlying cost differences. The basic arguments for price discrimination in the presence of decreasing costs and other than inelastic demand have been made in a number of places[7] and will only be briefly summarized here. Marginal cost pricing in decreasing-cost industries still preserves its optimality properties but fails to cover costs. Charging at what the traffic will bear in selected services can make up that deficit. Indeed, the production economies realized by attracting additional marginal, high-elasticity traffic at lower rates will rebound not only to the obvious benefit of that traffic but to users discriminated against, who will have lower costs to pay than would otherwise have been the case.

In a theoretical setting, Kahn describes "perfect" price discrimination, which "involves fashioning charges according to what each unit of traffic will bear. Its basis is differences not in cost but in demand, in the value of each unit of service to each purchaser."[8] One job of the regulatory authorities is to prevent this from occurring when it would mean above-normal profits to the monopolistic supplier of a particular important service. More practically, price discrimination can be

> the means by which public utility companies can (1) cover (or come close to covering) average total costs, while (2) making fuller use of existing capacity and/or (3) taking fuller advantage of long-run decreasing costs, by (4) permitting as many purchases are possible to be made as long as buyers are willing to pay the incremental costs of supplying them.[9]

Particular rules have been suggested for implementing price-discriminatory tariff structures. For example, Kahn urges that no prices, discriminatory or otherwise, be set below marginal or above average total costs: otherwise, "internal subsidization" or cross-subsidization of services is said to occur.[10] Furthermore, the only equitable way of deciding which traffic is "marginal" and which traffic is "inframarginal," in terms of who enjoys the benefits of decreasing costs derived from larger production, is to rank users by elasticity of demand in the relevant range, beginning with the most inelastic. (The subsidization criterion has been used above regarding regional pricing; and for costing purposes NASA traffic was assumed inframarginal or relatively inelastic to other traffic.) As to the particular levels of prices, one standard is that the prices of various services be set higher than marginal cost in inverse proportion to their respective demand elasticities. Baumol,[11] beginning from the case in which all prices equal average total costs, suggests that prices rise for inelastic users and fall for elastic users until the ratio (P-MC) ÷ MC is inversely proportional to the respective demand elasticities. And in perhaps the most rigorous and convincing defense of price discrimination as an adjunct to marginal cost pricing in the presence of decreasing costs, Baumol and Bradford, as noted in Chapter 3, argued that they had proposed a Pareto-optimal solution in the second-best situation of decreasing costs in this latter approach. In seems clear, on the basis of the arguments summarized and cited above, that price discrimination, far from being distortive and misallocative in all situations,[12] has quite desirable properties in situations of decreasing costs.

All the above arguments are of course vitiated by the extent to which costs are nondecreasing or demand is inelastic. This study has established pervasive decreasing costs in INTELSAT, so no problem in that regard exists for satellite communications. Yet the inelasticity of demand for satellite circuits has also been noted. INTELSAT, for example, projects traffic with no assumptions regarding demand elasticity. The chain of marketing of telecommunications satellites from the INTELSAT space segment through the INTELSAT signatory and into the domestic distribution networks involves a good many price markups and subsequent services, so that the final consumer may well be insensitive even to a large increase in the price of what constitutes only a small part of the total service he is actually purchasing. Clearly, however, at least for low ranges of output, there must be some price elasticity, if only in the sense that there must be some ceiling price which would choke off all demand. Furthermore, at low price levels, there must be at least some traffic which would not otherwise be forthcoming. Little is known about demand elasticity for satellite circuits, and this would appear to be a matter of priority for further research.

As Table 4.9 indicated, the only areas in which value-to-user pricing has entered the INTELSAT tariff to any extent are in the various multidestination charges (other than for telegraph bearers), and in recent transponder and half-transponder lease decisions. There is an additional area for possible value-of-service charging in the INTELSAT system, namely that of distance-based pricing.

To begin with, it is clear that INTELSAT costs are almost entirely insensitive to the terrestrial distance between communicating earth stations. A counterexample might be narrowly-focused spot beams, which provide lower-cost communications within a small area of coverage. But, as noted, the use of spot beams is not really at the discretion of individual purchasers, and it changes from time to time according to overall criteria of system optimality.

The distance-inelastic feature of satellite communications costs is presumably also shared by high-frequency radio waves, which, however, are a virtually noncompetitive mode by now. The costs of cables, the main competing mode of terrestrial and overseas communication, are clearly distance-sensitive. Thus, one factor affecting the relative price elasticities of demand on individual satellite links will be the presence or absence of cable connections, much like the presence or absence of a navigable river alongside a railroad affects the demand schedule the railroad faces by making demand more elastic than it would be otherwise.

Application of link-specific tariffs based on demand elasticities might produce some anomalies in practice. ("Distance-based charging" is a misleading term: distance is only one factor in determining demand elasticity.) For example, North Atlantic links, such as between New York and London (more correctly, between the earth stations in Andover, Maine, and Goonhilly Downs, Cornwall), would probably have quite low tariffs due to the presence of numerous competing cables. Shorter links in the absence of cables, such as between individual Latin American or African countries separated by jungle or mountains, might have higher rates. Again, they might not if there is little potential traffic between such countries. Ceteris paribus, however, one would expect demand-based, link-specific satellite rates to increase with distance.

Unlike regional pricing, which seemed to imply at least some subsidization of less developed countries by industrialized nations, link-specific charging would probably have the opposite effect. Cables join primarily richer, industrialized countries. Small countries having invested millions of dollars to join INTELSAT and to build earth stations, however, might not sit idly by as their rates were doubled or tripled. They could argue convincingly that they had joined INTELSAT with every assurance that the well-established global pricing policy would continue. Article IV(d) of the INTELSAT Agreement is on their side as well. Few industrialized countries would take the political risks to argue for link-specific pricing by interpreting the type of utilization as varying from link to link. Most would agree that there is only one major type of utilization, that is a unit of utilization beamed to and received from an in-orbit satellite, the origin and destination not reflecting cost differences and therefore being irrelevant. Recall that a similar problem of interpretation arose above with respect to regional pricing. The increased politicization of INTELSAT's structure under the permanent arrangements, noted in Chapter 1, rebounds to the benefit of smaller member countries and would further militate against any interpretation of Article IV(d) against their interests.

Perhaps, however, it might turn out that high-traffic links between industrialized countries, even in the presence of cables, are more demand-inelastic in certain cases than are low-traffic links between small countries with no cable alternatives. Only in this situation, assuming all the countries could agree either to a convenient interpretation of Article IV(d) or to its amendment or repeal, would the political and institutional drawbacks of a link-specific charge scheme based on price discrimination appear small enough to warrant its implementation.[13]

NOTES

1. See Marcellus S. Snow, "Investment Cost Minimization for Communications Satellite Capacity: Refinement and Application of the Chenery-Manne-Srinivasan Model," Bell Journal of Economics 6, no. 2 (Autumn 1975); this contains a reference to the nonoptimality of investment behavior occasioned by the fact that the same satellites are often launched in different ocean regions with different traffic requirements and demand growth rates. See also Kenneth Bishop Stanley, "International Telecommunications Industry: Interdependence of Market Structure and Performance Under Regulation," Land Economics 49, no. 4 (November 1973), pp. 396-97. It is of note that in considering follow-on systems to the INTELSAT IV series, a separate satellite was considered (but not subsequently adopted) for the Indian Ocean region: see ICSC-56-21.

2. Alfred E. Kahn, The Economics of Regulation, vol. 1, Principles (New York: John Wiley, 1970), pp. 142-43.

3. INTELSAT Agreement, Article III(a), p. 9.

4. Ibid., Article IV(d), p. 13.

5. Ibid.

6. ICSC/F-23-3, March 8, 1971, p. 3.

7. Kahn, op. cit., chap. 5, is a comprehensive summary. Three earlier articles give diagrammatic treatments of price discrimination in the presence of decreasing average costs: Eli W. Clemens, "Price Discrimination in Decreasing Cost Industries," American Economic Review, 31, no. 4 (December 1941), pp. 794-802; D. A. Worcester, Jr., "Justifiable Price 'Discrimination' under Conditions of Natural Monopoly: A Diagrammatic Representation," American Economic Review 38, no. 3 (June 1948), pp. 382-99; Eli W. Clemens, "The Marginal Revenue Curve under Price Discrimination," American Economic Review 38, no. 3 (June 1948), p. 388-90.

8. Kahn, op. cit., p. 132.

9. Ibid., p. 133.

10. Ibid., pp. 141-43.

11. William J. Baumol, "Reasonable Rules for Rate Regulation," in Prices: Issues in Theory, Practice, and Public Policy, eds., A. A. Phillips and Oliver E. Williamson (Philadelphia: University of Pennsylvania Press 1967), pp. 122-23.

12. See Stanley, op. cit., pp. 400-01. Stanley seems to indict the price discrimination in Comsat's rate structure as a cause of "resource misallocation" without a discussion of its optimality properties. It is not enough to dismiss price discrimination as being based on "cable costs and an ad hoc reaction to cable applications." It is cable costs and cable prices that have a strong effect on the price elasticity of demand for satellite circuits, and it is that price elasticity, rather than underlying modal technology, that most powerfully determines the optimality properties of particular changing policies for satellite service.

13. See Stanley, op. cit., p. 400. While INTELSAT has never adopted distance-based charging, it is noteworthy that Comsat has done so in its wholesaling of INTELSAT circuits to the U.S. international carriers. For example, in 1973 a voice circuit from the U.S. mainland to Japan was tariffed at $4,900 per month, while comparable service to Puerto Rico cost the carriers $1,500 per month.

PART III

ADDITIONAL ECONOMIC ISSUES

CHAPTER

6

SEPARATE SYSTEMS: THE QUESTION OF ENTRY

The purpose of Part III is to survey briefly some additional economic issues relating to the experience of INTELSAT. None of these issues has been dealt with to date in a thorough way from an economic standpoint, and all would appear to invite further scholarly investigation. Chapters 6-8 will outline economic questions in a number of separate areas such as regional systems and development via telecommunications. The interested reader is referred to the notes in Part III for further references, as well as to the extensive bibliographies in studies dealing with INTELSAT and satellite communications from the standpoint of international politics.[1]

Chapter 2 of the present study established and examined pervasive declining average costs in INTELSAT's operation and technology. It is this characteristic which is the crux of the so-called natural monopoly argument used to justify the restriction of entry of additional firms to a given activity or industry. There are additional natural monopoly arguments that would also appear applicable to international satellite communications. The most important of these involves the negative production externalities involved in the use of certain scarce resources, namely the radio frequency spectrum and the number of orbital parking spaces, which might be adversely affected by unregulated competition.

Findings supporting the concept of a single global system were also reported in the Rostow Task Force Report on international telecommunications, issued by the U.S. government in 1968:

> The global system enables substantial economies of scale to be realized. This is true because large satellites are capable of flexibility in use and provide high communications capacity and lower costs per channel. They also permit economies in the use of earth stations. This fact is of special importance to the developing countries, whose communications systems are often still limited, costly and indirect. There are other

> important advantages to a global system in integrated system planning, financing, procurement, management, and control.[2]

For somewhat different reasons, a White Paper on the possibility of a Canadian domestic system also found a satellite system to be a natural monopoly: in many parts of Canada other than the south, there is "no effective alternative."[3]

If one accepts the claim that international satellite communications constitutes a natural monopoly in the usual sense of the word, one must also accept the proposition that, assuming enlightened regulation of the monopolist, costs would be minimized, and therefore economic welfare maximized in some meaningful sense, if only a single global satellite system provided such services. This statement, furthermore, is applicable to all sources of satellite communications, domestic or international, public or even military. It assumes homogeneity of the service provided to the extent that variations in service requirements could be met on large satellites without loss of scale economies. Under that criterion, many noncommunications applications of satellites, such as earth resources, surveys, mappings, and so forth, could not be included in the natural monopoly argument.

For obvious political reasons, however, there is no possibility that all world telecommunications traffic can or will ever be carried on the same system of satellites. First of all, national security considerations would prevent mingling of military with civilian traffic, either of the same country or particularly of different countries. Second, military satellite technology is far more advanced than civilian satellite technology, and separate military systems already exist. As has been noted previously, there may well have been subsidization of civilian satellite communications in that Hughes and TRW Systems, the two prime INTELSAT contractors, had built military satellites before they built INTELSAT satellites, and some of the earlier technology and managment skills may well have influenced the design of the later models without being wholly reflected in contract prices.[4] These substantial spillovers must affect any conclusions reached regarding costs and prices of later civilian systems. In any event, carriage of military traffic is prohibited by the INTELSAT Agreement[5] and is politely ignored at international conferences of civilian communicators:

> Quite often, in the United States and Britain, the launching of a military satellite is accompanied by a flurry of press releases, just as if it were a commerical satellite. But after that, silence. . . . To mention military satellites at the ITU or at a conference about the use of satellite communications for the good of mankind is a breach of good manners.[6]

In addition to individual countries, notably the United States, Great Britain,[7] and the Soviet Union, NATO has its own system of satellites for essentially military purposes. Since in the eyes of these nations the desirablility of military traffic being carried on separate satellite systems owned and operated by them outweighs the economic advantages to be gained from economies of scale in its carriage on a single international (or even a single national) system, military traffic will not be considered in the rest of this chapter.

That exclusion leaves a good deal more in the residual, however, than the "international public traffic" the carriage of which is set forth as INTELSAT's "prime objective."[8] In particular, the INTELSAT Agreement provides specifically for the carriage of "domestic traffic" and "special services" in addition to international public traffic; the relevant language is in Article XIV.

Indeed, the requirements for separate systems serving "domestic public telecommunications" and "specialized telecommunications services, domestic or international" are not restrictive despite INTELSAT's economic interest in carrying all such traffic. All that is required for the provision of such separate systems is that INTELSAT's Board of Governors be provided with certain types of information. For separate domestic systems, INTELSAT members are to "consult" the Board of Governors, which in turn is required to

> express, in the form of recommendations, its findings regarding the technical compatibility of such facilities and their operation with the use of the radio frequency spectrum and orbital space by the existing or planned INTELSAT space segment.[9]

An INTELSAT member desiring to acquire or use satellite facilities other than those of INTELSAT for specialized services, international or domestic, is merely required to "furnish all relevant information to the Assembly of Parties, through the Board of Governors."[10]

For non-INTELSAT systems in the above two categories, then, INTELSAT members are merely required to furnish information of a technical nature, on the basis of which appropriate INTELSAT bodies make recommendations regarding technical and operational compatibility, and so forth. No mention is made of economic compatibility with INTELSAT or the possible draining off of INTELSAT traffic. By contrast, INTELSAT members contemplating establishment or use of separate facilities for public internation traffic, that is, the same kind of traffic INTELSAT is obliged to carry as its prime objective, are under much stricter obligations from the INTELSAT Agreement:

> To the extent that any Party or Signatory or person within the jurisdiction of a Party intends individually or jointly to establish, acquire or utilize space segment facilities separate from the INTELSAT space segment facilities to meet its <u>international public telecommunications</u> services requirements,

> [emphasis supplied], such Party or Signatory, prior to the establishment, acquisition, or utilization of such facilities, shall furnish all relevant information to and shall consult with the Assembly of Parties, through the Board of Governors, to ensure technical compatibility of such facilities and their operation with the use of the radio frequency spectrum and orbital space by the existing or planned INTELSAT space segment and to avoid significant economic harm to the global system of INTELSAT [emphasis supplied].[11]

The avoidance of "significant economic harm" as a prerequisite for INTELSAT's approval of a separate system for public international traffic was the object of heated debate during the negotiations of the permanent agreement. Most European countries (particularly France, with ambitions for a continuation of its rayonnement culturel to its former colonies and other French-speaking areas, such as Quebec, via a separate satellite system) wished no strictures whatever to be placed on the establishment of separate systems. The U.S. government and Comsat were for obvious reasons anxious for restrictions against separate systems causing any adverse economic impact whatever on INTELSAT. The developing nations were torn between their desire for the cheapest possible communications and their political and cultural ties to Europe. In the end, the determination of significant economic harm as the criterion was a compromise; this partial victory for the United States was due at least in part to the unexpressed but potent threat of NASA's refusal to launch the satellites of any system felt to be inimical to INTELSAT's economic interests.

What forces have been at work to encourage the carriage by INTELSAT of domestic and special traffic with the assumed benefits of scale economies accruing to all users? What, on the other hand, have been the centrifugal forces contributing to the conception and, in some cases, establishment, of complementary and sometimes competing systems? Specific satellite projects which illustrate both tendencies are outlined below.

EUROPEAN REGIONAL SYSTEM

Since the early days of INTELSAT operations, plans have been afoot among European countries to establish a separate European regional satellite system. Most of the planning and initiative for such a system has come from the European Space Research Organization (ESRO). In its various conceptions, the system was to include some intra-European traffic (for example, long links such as from Cyprus to Ireland for which satellites would be economically competitive with existing terrestrial facilities), as well as telephone and television traffic between Europe and Africa. At first, the intent of the Europeans was to launch such a system with European rockets, which the European Launcher Development Organization (ELDO) was assigned to develop. ELDO was plagued, however, with a disappointing series of 11 straight launch failures[12] in an effort to perfect its Europa-1 and Europa-2 rockets. In some instances the rockets used were multistage, each of the stages being furnished by a major European country. As the availability of a European launcher became more and more dubious by the late 1960s, the terms on which NASA (the only other potential source of launch services besides the Soviet Union) would provide launchers for a separate European satellite system became, as noted, a bargaining object in the INTELSAT permanent arrangements negotiations which began in 1969.[13] NASA has since proclaimed its readiness to launch essentially any civilian satellite for foreign countries, but this declaration came only after the INTELSAT agreements entered into force in 1973 with the "significant economic harm" provision referred to above. ESRO and ELDO, now in the process of merging into a single European Space Agency,[14] still plan a separate European system by the end of the 1970s, presumably to be launched by American rockets.[15] The European system can be seen as a costly but ultimately successful effort to achieve a degree of independence from America's domination of space technology and its real or imagined hegemony in INTELSAT. A side effect has been to bolster the European space and telecommunications industry, which had not obtained the hoped-for degree of participation in INTELSAT contracting (see Chapter 8). The expense to European countries in terms of the higher unit costs of the system ultimately to be launched (costs which must include the unsuccessful attempt to develop a separate European launcher) was presumably judged less, by the countries involved, than the benefits in terms of the prestige, independence, and flexibility offered by a separate regional satellite system. This is not to imply that European views were uniformly in favor of all aspects of the system. Great Britain, for example, was strongly opposed to the launcher development program after its costs mounted and its ultimate failure became a distinct possibility, and withdrew its financial support for the launcher program in 1970.

DOMESTIC SATELLITE SYSTEMS

The U.S. Domestic System

As early as 1966 Comsat asked the FCC for permission to establish a domestic U.S. satellite system, and the 1968 Rostow Task Force

Report[16] backed Comsat for the job. At the time, Comsat had approximately $200 million in surplus capital, the global system having cost much less than originally planned, and it was earning more money on bank deposits than on revenues from INTELSAT circuits it wholesaled to the U.S. carriers. Comsat was anxious to expand its operations and revenues, as well as its rate base, by establishing a domestic system, including both satellites and earth stations.

These hopes were dealt a setback in 1970 when the White House issued a memorandum[17] calling for a policy of allowing several entrants into the domestic market rather than granting a monopoly to Comsat or another applicant. Claiming that "no natural monopoly conditions appear to exist in the provision of specialized communications via satellite," the memorandum proposed to allow

> competition [in providing a domestic satellite system] to act within well-defined limits necessary to preclude anti-competitive practices and to assure that the competition works toward the public interest.[18]

On the basis of that procompetitive recommendation, the FCC has approved a number of applications for provision of a domestic system.[19] A system operated by Western Union began commercial operation in July 1974.[20] Proposals by Comsat and AT&T have also been approved but are farther from implementation. It should be noted that the claim of absence of scale economies made by the White House memorandum was accompanied by no supporting analysis and goes counter to the findings of the present investigation as well as the Rostow Task Force Report.

The entry of competing firms into the U.S. domestic satellite market should provide a powerful test of the hypothesis that that market is a natural monopoly. If it is, economic theory would predict the emergence of destructive competition among entrants, such as rate wars in which prices charged (or at least applied for from the FCC) would fall to marginal cost, causing short-term operating deficits because of decreasing average costs. These costs would presumably be borne by entrants as a short-term strategy to establish a larger market share, reap economies of larger production, and drive out competitors in the long run. If, alternatively, decreasing costs were limited relative to the size of the market, competition among several entrants would presumably not be "destructive," in the sense that each entrant could charge prices near marginal costs while covering average costs and maintaining its market share. Only in the former case, it would appear, is extensive government regulation necessary or desirable.

Canadian Domestic System

A Canadian domestic system was proposed in a White Paper of the Candian government issued in 1968.[21] Initial hopes for a Canadian prime contract were disappointed when Huges Aircraft issued the lowest bid, which was accepted. Subsequently, following an initial launch by NASA in late 1972, the Canadian domestic system entered its early operational phases.[22]

INTERSPUTNIK—THE SOVIET INTERNATIONAL SYSTEM

The Soviet Union was the first nation to operate a domestic satellite system. That system, named Orbita and based on random-orbiting Molniya satellites, was first established in 1966. In 1968, the Soviet Union and seven other socialist countries submitted a draft agreement to the United Nations establishing INTERSPUTNIK, an international communications satellite organization.[23] Its main deviations from the then existing interim INTELSAT model included legal personality (subsequently adopted in the permanent arrangements by INTELSAT) and a one-man, one-vote governing body whose decisions were to be taken by a two-thirds majority.[24]

Nine states finally signed a revised agreement for INTERSPUTNIK at Moscow on November 15, 1971. That agreement made clear that the Soviet Union envisaged an evolution of the Soviet domestic system into the international system, joining it to its allies. In the first of three stages, existing Soviet in-orbit capacity was to be provided free of charge to INTERSPUTNIK members. After the end of 1973, an operational phase was to begin in which channels were to be leased from the Soviet Union, again presumably on spare capacity from the existing Soviet domestic system. A final phase envisages the construction, launch, and lease of separate INTERSPUTNIK capacity by members of that organization.[25] Launch, monitoring, and control functions are to be provided bilaterally between INTERSPUTNIK and "those member States possessing the necessary facilities." Channels are to be made available "at fixed tariffs on a par with the average world tariffs for international telephone channels"; eventually, one or two geostationary satellites of the Soviet Statsionar series are to replace the random-orbiting Molniya satellites.[26] Little is known about the present status of INTERSPUTNIK, and the Soviet Union appears to be following its usual practice of withholding information about space activities until solid successes can be reported.

Although the Soviet Union and its allies were invited to join INTELSAT when it was first organized, they declined to do so. The Soviet motive for refusal was officially expressed as U.S. domination of the system. Another reason was clearly the voting and investment arrangements based on usage, which would have given the Soviet Union only 2 or 3 percent of the system's vote and ownership, placing it, for example, behind Switzerland. Technological backwardness has also been suggested[27] despite the Soviet Union's early successes in subsynchronous rocketry. Presently Yugoslavia is the only Communist country belonging to INTELSAT. The People's Republic of China, as noted earlier, has been a customer of INTELSAT from time to time, notably during former President Nixon's trip to China.

SYMPHONIE PROJECT

The bilateral Symphonie project between France and West Germany is more experimental than operational in nature; nonetheless, plans for the system have been carefully coordinated with competent INTELSAT bodies. The Symphonie project dates from 1967, when a joint agreement called for "the design, construction, launching, and utilization of an experimental telecommunications satellite for the purpose of distributing radio and television broadcasts, providing telephone and telegraph communications, and transmitting data."[28] There is a meticulously equal sharing of costs, decision-making, personnel, and contracting. Use of earth stations with 15-meter antennas as well as mobile stations of 6-8 meters in diameter (both nonstandard in the INTELSAT sense) is planned.

Original plans called for launching of the Symphonie satellites from the Kourou base in French Guiana, using a Europa-2 launcher. Because of the failure of the ELDO rocket program, however, an agreement was reached with NASA calling for that agency to launch the first flight model in late 1974 or early 1975.[29] The first launch by NASA was in January 1975.

SPECIAL SYSTEMS: AERONAUTICAL AND MARITIME

Satellites have special promise for providing navigation for ships at sea and aircraft crossing the high seas. Institutional and technical problems have been primarily responsible for delays in implementing satellite technology in these two areas.

Aeronautical Satellite Systems

No purpose would be served here in recounting in detail the long and convoluted history of proposals and negotiations, dating from 1966, regarding the establishment of satellite systems for aircraft communications and position-fixing in the Atlantic and Pacific regions.[30] INTELSAT played a role in early consideration, but opposition, particularly from the French, who had conducted significant research on new avionic technology, ultimately left INTELSAT out of the system. Airline interests were opposed to a new technology and adoption of a different, higher frequency because of the extra on-board costs that would be incurred by each aircraft. In mid-1974, ESRO, the moving force in Europe for an aeronautical satellite system, signed a memorandum of understanding with Canada and the United States for the establishment of an Atlantic system with the Canadian government and a U. S. industrial partner (Comsat). The U. S. share of system development costs, estimated at $100 million, is to be paid out of the Airport and Airways Trust Fund instead of by the airlines. There will be two satellites in the Atlantic system, the first to be launched in late 1977. Spacecraft will have both the present (VHF) frequency, and the higher, experimental (L-band or UHF) frequency contemplated for future use. The jointly owned system will include the space segment and ground stations. An Aerostat space segment program office is now functioning in the Netherlands and will issue specifications for the satellites.[31]

Perhaps the salient feature of these developments for present purposes is the eventual organization of the aeronautical system outside the framework of INTELSAT. This reflects substantial sentiment in that body that INTELSAT's main task is to serve public telecommunications traffic rather than specialized areas such as avionics. As noted, that view is reflected by the lenient language in Article XIV(c) of the INTELSAT Agreement regarding specialized services.

Maritime Satellite Systems

Initiatives for maritime satellite systems from three separate directions have been undertaken, each in a different stage of development.

Comsat's Maritime System (MARISAT)

Through its wholly-owned subsidiary, Comsat General, Comsat has organizaed a maritime satellite system (MARISAT) to be in service in the Atlantic and Pacific by the mid-1970s. Initially, the biggest user will be the U.S. Navy; after a three-year period, commercial services including telephone, telex, data, and facsimile will provide the main source of system revenues. Comsat General is to own about 80 percent of the system, the rest being shared by the U.S. international carriers. There will be one satellite over each ocean (Atlantic and Pacific) with a third on-ground spare.[32]

The ESRO MAROTS Maritime Satellite System

Using the OTS (Orbital Test Satellite) of the ESRO space program, the so-called MAROTS maritime system is scheduled for launch and operation in late 1977 under ESRO's aegis. There will be one Atlantic satellite and one on-ground spare.[33]

INTELSAT Maritime System

INTELSAT consideration in the maritime area dates from late 1972 and is still in the study phase.[34] In fact, the Finance Subcommittee has presented the Board of Governors with a study of the financial impact of an INTELSAT maritime satellite system.[35] The Board of Governors Finance Subcommittee determined both incremental and fully allocated costs for the system assuming modification of INTELSAT V satellites to accommodate such specialized traffic.[36] As in its earlier analysis of whole transponder leasing, it hinted at value-of-service as a possible criterion of charge.[37]

SEPARATE SYSTEMS: CONCLUSIONS

The survey above has indicated a number of areas in which satellite systems, whether for domestic, international, experimental, or specialized purposes, are being and have been established outside the INTELSAT framework, or are presently planned or contemplated. Yet it must be concluded that in relation to the size of INTELSAT in both traffic and capital, these efforts are modest. INTELSAT has established itself as the leading global communications satellite system more firmly than its birth pains and the acrimony of its early years and renegotiation gave reason to hope. This is confirmed not only by the limited extent of competing or complementary systems, but more particularly by the growing list of countries which have leased whole or half-transponders on INTELSAT capacity.

It seems clear that motives other than economic ones have prompted entry of separate systems into INTELSAT's former near monopoly. For larger nations, such as the Soviet Union, the United States, and Canada, such motives are to some extent political in nature. In addition, perhaps traffic in these countries is great enough to realize scale economies of its own close to those of INTELSAT. The European satellite project, while strongly political in its own right, is probably also inspired as an engine of technological self-sufficiency for Western Europe vis-a-vis the United States. The Symphonie system, as a special case, is experimental in nature but may carry some commercial traffic;[38] considerations of politics as well as technological advance and industrial contracting figured in this project. Specialized systems in the aeronautical and maritime areas, by contrast, may well be cheaper under special arrangements outside the INTELSAT framework, being designed for traffic substantially different from that carried by INTELSAT.

NOTES

1. Joseph N. Pelton, Global Communciations Satellite Policy: INTELSAT, Politics and Functionalism (Mt. Airy, Md.: Lomond Books, 1974); Judith T. Kildow, INTELSAT: Policy-Maker's Dilemma (Lexington, Mass.: Lexington Books, 1973); Jonathan F. Galloway, The Politics and Technology of Satellite Communication (Lexington, Mass.: Lexington Books, 1972). The latter is especially complete in references to public documents.
2. U.S. Office of the President, "Report of the Task Force on Communications Policy," December 7, 1968, chap. 3, p. 12, quoted in Kildow, op. cit., p. 62.
3. Canada, Ministry of Industry, White Paper on a Domestic Satellite Communication System for Canada, March 28, 1968, by C. M. Drury, minister of industry, para. 20, p. 58.
4. See Brenda Maddox, Beyond Babel: New Directions in Communications (New York: Simon and Schuster, 1972), pp. 79-80.
5. See specific prohibitions in the INTELSAT Agreement, Article III(d) and III(e)(iii), p. 10.
6. Maddox, op. cit., p. 79.
7. For a description of Skynet, the British military satellite system, see N. Simmons, "The United Kingdom Program of Communication Satellites," Communications Satellite Systems, ed. P. L. Bargellini (Cambridge, Mass.: M.I.T. Press, 1974).
8. INTELSAT Agreement, Article III(a), p. 9.
9. Ibid., Article XIV(c), pp. 43-44.
10. Ibid., Article XIV(e), p. 44.
11. Ibid., Article XIV(d), p. 44.
12. Pelton, op. cit., p. 81.
13. See Kildow, op. cit., p. 75.
14. For details on difficulties involved in the merger, see Donald C. Winston, "European Space Agency Shift Stalled," Aviation Week & Space Technology, April 22, 1974, p. 19.

15. For details see Communications Satellite Corporation, Geneva Office Handbook (October 1973), pp. 314-19; and Rene C. Collette, "The European Communication Satellite Program," in Bargellini, op. cit.

16. U.S., President's Task Force on Communications Policy, Final Report, December 7, 1968, chap. 5.

17. U.S., Office of the President, "Memorandum for the Honorable Dean Burch, Chairman of the Federal Communications Commission," January 23, 1970.

18. Ibid.

19. Katherine Johnsen, "Four Domestic Satcom Plans Authorized," Aviation Week & Space Technology, September 17, 1973, p. 18.

20. See New York Times, April 15, 1974. Western Union launched the first satellite of its Westar domestic system on April 13, 1974; the system will cost $90 million, including 5 earth stations and 3 satellites to be boosted by Delta rockets. Earth stations will be located in New York, Atlanta, Chicago, Dallas, and Los Angeles. See also New York Times, January 9, 1974, and Telecommunications Reports, January 14, 1974; in January 1974, RCA, using capacity on the Canadian domestic satellite Anik II, demonstrated a domestic satellite system linking Alaska with the lower 48 states. On September 18, 1973, Comsat awarded a $65.9 million contract to Hughes Aircraft to build four high-capacity domestic satellites from which circuits will be leased to AT&T. Three are planned in orbit with a fourth as a spare; see New York Times, September 19, 1973. As of October, 1973, NASA had received down payments to provide launching services for "at least 16" domestic communications satellites over the next two years; see New York Times, October 1, 1973. Late in 1974, the FCC rejected a bid by IBM and Comsat to establish an advanced domestic system on a 55 percent/ 45 percent ownership basis; see "FCC Withholds Approval," Aviation Week & Space Technology, February 3, 1975, p. 17.

21. Canada, Ministry of Industry, op. cit. See also Charles M. Dalfen, "The Telesat Canada Domestic Communications Satellite System," Stanford Journal of International Studies 5 (June 1970), pp. 84-113.

22. For recent details see B. C. Blevis and M. L. Card, "The Implications of Satellite Technology for Television Broadcasting in Canada," in Bargellini, op. cit; and Telecommunications Reports, March 11, 1974, p. 22.

23. United Nations, Document A/AD.105/46, August 9, 1968. For a detailed analysis of this formative period see Galloway, op. cit., chap. 7; and Thomas L. Shillinglaw, "The Soviet Union and International Satellite Telecommunications," Stanford Journal of International Studies 5 (June 1970), pp. 199-226.

24. Communications Satellite Corporation, op. cit., October, 1972, p. 154.

25. Ibid., October 1973, p. 302.

26. Ibid., pp. 302-03.

27. Pelton, op. cit., p. 55.

28. Symphonie Agreement, quoted in Communications Satellite Corporation, op. cit., October 1973, p. 326.

29. Ibid., pp. 328-29. See also F. Hiergeist, J. L. DeMontlivault, and G. Payet, "The Earth Segment in the Symphonie Project," in Bargellini, op. cit.; and "Symphonie May Provide Commercial Service," Aviation Week & Space Technology, September 30, 1974, p. 22.

30. See Communications Satellite Corporation, op. cit., October 1973, pp. 305-11, for a competent summary through mid-1973.

31. "Aerosat Program Near Final Agreement," Aviation Week & Space Technology, May 13, 1974, p. 15. For more recent details, see "U.S. Firms to Vie for Aerosat Award," loc. cit., December 16, 1974, p. 24.

32. Communications Satellite Corporation, op. cit., October 1973, pp. 320-21. For additional detail see Stephen E. Scrupski, "Marisat Partners Seek Accord," Electronics, June 13, 1974.

33. Communications Satellite Corporation, op. cit., October 1973, p. 321. For contract awards see Telecommunication Journal, May 1974.

34. Communications Satellite Corporation, op. cit., October 1973, p. 322.

35. BG/F-5-3, March 17, 1974, paras. 21-38, pp. 10-18.

36. Ibid., paras. 25-34, pp. 12-16.

37. Ibid., para. 25, p. 12.

38. See "Symphonie May Provide Commercial Service," Aviation Week & Space Technology, September 30, 1974, p. 22.

CHAPTER

7

NATURAL RESOURCES AND ECONOMIC DEVELOPMENT

NATURAL RESOURCE IMPLICATIONS OF SATELLITE COMMUNICATIONS

Very few Congressional hearings into satellite communications end without an obligatory reference to the radio frequency spectrum as a "scarce natural resource" that must be "preserved." Yet the implied analogy with depletable resources such as fossil fuels is not wholly apt. In the first place, the spectrum is a resource which is "used, not consumed. . . . [It] is being wasted when it is not being used."[1] It is dimensioned differently from other natural resources: it is not a stock variable but rather has dimensions of "space, time, and frequency," all of which are interrelated.[2] It is not depletable; once used, it can be used again. Furthermore, the same frequencies can be used at the same time if the geographical separation is far enough to prevent interference. The spectrum is, of course, subject to a kind of pollution by interference caused by overcrowding.[3]

The ITU is the international organization charged with regulating the use of the frequency spectrum. Its subsidiary body, the International Frequency Registration Board (IFRB), keeps a master list of all used and apportioned parts of the spectrum, but neither body has powers of enforcement beyond those of moral suasion. The INTELSAT system uses frequencies which were specifically allocated for satellites at two ITU-sponsored conferences in 1963 and 1971.[4] These are primarily in the range of the so-called microwaves, from 3 to 30 gHz. Shorter wavelengths, the "millimeter waves" from 30 to 300 gHz, are also allocated but at present cannot be used because they are so short that they are attenuated and rendered useless by atmospheric precipitation. Research is underway in this area, however, and its expected ultimate success will be an example of how economic pressures can lead to an expansion of natural resource possibilities.

A second, more problematical and controversial resource issue is posed by the number of so-called orbital parking slots for synchronous satellites. The circumference of the band stretching 22, 300 miles above the equator and about the earth is approximately 170, 000 miles, and it is along or near that band that all present and future stationary satellites must presumably be located. This would appear to be a truly one-dimensioned, stock resource capable of depletion in the normal sense, at least to the extent defunct satellites cannot be physically removed.

The chance of satellites actually colliding along this band is almost nil and will be ignored in the following discussion. The much more significant danger, according to many observers, is that as stationary satellites become numerous they will begin to interfere with each other's signals. The Task Force on International Satellite Communications, however, financed by the Twentieth Century Fund, discounted the scarce-resource arguments relating to orbital parking slots along the following lines: satellites can now be controlled so as not to stray more than one-tenth of one degree in either direction from their orbital space. Allowing two-tenths of one degree for each synchronous satellite means that 1,800 satellites can be accommodated along the 170,000-mile stationary band, leaving them slighly less than 100 miles apart. The task force concluded that satellites this close could still prevent interference using spot beams narrowly focused on different areas, as well as special techniques such as orthogonal polarization of antennas, frequency reversal, and so on.[5]

A 1970 conference of experts called by the Twentieth Century Fund to consider these findings was divided in its reaction to this discounting of the scarcity of orbital parking slots. Fears were expressed that the desirable places were already being seized by the countries with launch capabilities and that "if the theory of abundance later turns out to be wrong, then the countries that came later to the feast will suffer for it."[6]

The lack of regulation of the kind applied to the frequency spectrum aroused fears of chaotic situations like many in the past which had arisen because of poor frequency planning. The conference concluded:

> One basic problem . . . was that satellites are being placed in orbit without regard for long-range global considerations; no organization is in charge of regulating or even studying the question of spacing in orbit.
>
> All agreed on the need for planning. And when all the opposing views had been aired, the gulf between those who regarded the geostationary orbit as abundant and those who feared that it was scarce had narrowed considerably. There was a consensus that those who used the orbit should be required, periodically, to account for their use to some organization which would be able to keep track . . . of the technological developments affecting the capacity of the geostationary orbit. The most important thing . . . was that each user design his system so that it preserved the rights of others.

> The group . . . saw that the objective of maintaining flexibility was inconsistent with the notion of preassigned slots. Everyone was in favor of constant revision to make sure the orbit was being well used. No one wanted an arrangement which would leave latecomers out in the cold.[7]

This divergence of views characterizes rather well the present state of affairs regarding planning of the geostationary orbit. While there is a commitment in principle to the need for planning and coordination, fear of the congestion and depletion of this resource is not intense enough to prompt efforts at record-keeping and moral suasion such as those applied by the ITU and the IFRB with respect to the radio frequency spectrum. The White House memorandum of early 1970 dealing with the U.S. domestic satellite system, for example, discounted the scarce-resource arguments applied to both the spectrum and the geostationary orbit:

> The issue of radio resource scarcity for satellite communications has been overstated to a significant degree. While the communications capacity of this resource is finite, the ability to accommodate additional radio services is greatly expandable through administrative, technological, and operational innovation. Both earth station and satellite design standards can be varied to assure adequate orbital capacity for both immediate requirements and likely near-term growth. Long-term growth can be accommodated through further refinement or additional frequency allocations, whichever is most economic.
>
> Since some of the orbital locations and associated spectrum usage of interest for United States domestic satellites might also be potentially useful to other western hemisphere nations, a question of United States monopolization could conceivably arise. However, even 10 to 12 United States domestic satellites . . . would represent only a small fraction of the number which could be accommodated for western hemisphere use with the current state of the art. Therefore, orbital capacity is not expected to be a problem at this time. As demand for satellite communication expands, it may become necessary to evolve additional international coordinating mechanisms; but this would likely involve the establishment of appropriate technical standards rather than the rationing of orbital positions.[8]

One cannot read this statement without at least a little uneasiness over the glibness with which potential problems are dismissed, particularly with respect to orbital parking space. The country acquiring squatter's rights would, of course, have an incentive to argue that there is no need for regulation and that there is room for all comers. It appears, then, that the dangers from exhaustion of the stock, depletable resource of orbital parking space, at present ample but wholly unregulated, are greater than the dangers of overuse or misuse of the radio frequency spectrum, for which at least some regulatory machinery exists and which ultimately is not depletable in the usual sense of the word.

SATELLITES AND ECONOMIC DEVELOPMENT

One searches the economics literature in vain for articles or books with titles like "Satellites and Economic Development." The bulk of scholarly work in any vaguely related sense might be best described under the heading of "Communications and Social-Political Development," an area at once differenct from and broader than that treated in the present study. Much of the emphasis of such literature[9] is on the software rather than hardware aspects of mass communications—particularly television for instructional and other purposes—and on the social and political rather than economic results to be attained (literacy, political education, health and sanitation, and so forth). A common criticism of INTELSAT is that it has remained a conventional communications supplier offering conventional communications by unconventional means.

It is undeniable that INTELSAT's structure and mission are not suited to the special needs of the developing world. For example, the emphasis on colossal standard earth stations, costing upwards of $3 million, makes it difficult for a small country to join. Perhaps a developing country would prefer several smaller antennas scattered about its hinterland; one large antenna near its capital city serves only the tiny fraction of its population with the need and means to make telephone calls to foreign countries, as well as the enclave of foreign businessmen, travelers, scholars, and journalists who wish to communicate with their home countries.

INTELSAT took a great step toward meeting the special needs of less developed countries by the decision to lease whole or half-transponders at special rates, as described in Chapter 4. Some of the countries which have purchased INTELSAT capacity under those special arrangements, such as Brazil, had earlier considered establishing their own domestic satellite systems at much greater expense. There can be no doubt that the discriminatory pricing of spare capacity in favor of these higher-elasticity users had induced them to use that capacity, prevented a proliferation of systems outside the INTELSAT framework, reduced prices to all INTELSAT users, and opened important new horizons to those countries in terms of their economic, social, and human development.

Whether capacity is purchased from INTELSAT or another source, however, satellites offer less developed areas the possibility of skipping over or accelerating some of the phases of development which earlier developers have had to go through, along the lines of Alexander Gerschenkron's thesis.[10] For example, satellite earth stations can be emplaced throughout a less developed country without the necessity of developing the dense terrestrial communications infrastructure of industrialized countries, although poor roads and lack of transport facilities may pose a problem. Satellites can also speed the development of human as well as physical capital. An Indian official has estimated that "illiteracy in India could be wiped out by 1981 through the use of satellites."[11] Elaborate plans by a number of countries have been frustrated by the high cost and unavailability of satellite capacity adapted to their particular needs, despite the concessionary rates offered by INTELSAT. Others, however, have gone ahead with plans for their own domestic or regional systems, notably Indonesia[12] and the Arab countries.[13]

Although India has not leased INTELSAT capacity in bulk, that country's involvement with satellites, though still in the planning stages, is a good example of development-oriented satellite usage. Since the late 1960s India has been promised the use of NASA applications technology satellites (ATS), but the availability date slipped several years in a row due to budget cuts at NASA and domestic U.S. uses of those satellites judged to be of higher priority. At last, the ATS-6 satellite, launched in early June 1974, is now available to India after having been used to televise health and training programs to rural areas in Appalachia, the Rocky Mountains, and Alaska. The ATS-6, however, will be a prelude to India's own satellite system, INSAT, which is scheduled for initial launch in 1977. The educational uses of these satellites are to include improved agricultural techniques, family planning and hygiene, school instruction, teacher education, and occupational skills.[14]

Satellites are particularly valuable for those developing countries which are large, have inaccessible hinterlands separated by jungle, water, mountains, or desert, and comprise ethnically and linguistically diverse populations. The next ten years should produce a number of examples of striking progress in countries that are able to make use of satellite systems for development purposes alongside conventional telecommunications.

Finally, mention should be made of the effect of the 1973 renegotiated INTELSAT agreements on the less developed countries. As noted before, the Assembly of Parties and Meeting of Signatories in the new INTELSAT structure are run on a one-man, one-vote basis and have thereby increased the voice of smaller users in the organization. In addition, Article IX of the agreement has numerous provisions making it possible for signatories with small traffic and investment quotas to combine to gain membership on the INTELSAT Board of Governors as well as provisions assuring regional representation on the board regardless of traffic levels.[15]

NOTES

1. David M. Leive, "Regulating the Uses of the Radio Spectrum," Stanford Journal of International Studies 5 (June 1970), p. 23.
2. Ibid., p. 23.
3. Ibid., p. 24.
4. Communications Satellite Corporation, Geneva Office Handbook, October 1973, p. 298.
5. Task Force on International Satellite Communications, The Future of Satellite Communications: Resource Management and the Needs of Nations (New York: Twentieth Century Fund, 1970), p. 17. See also Brenda Maddox, Beyond Babel: New Directions in Communications (New York: Simon and Schuster, 1972). Maddox mistakes two-tenths of a degree for two-tenths of a mile and claims that satellites can be as close together as two-tenths of a mile, that is, two city blocks apart. By this criterion, 850,000 satellites could be accommodated in synchronous orbit, a generous allowance indeed.
6. Planning for a Planet: An International Discussion on the Structure of Satellite Communications, Report on an International Conference at Cap Ferrat, France, Sponsored by the Carnegie Endowment for International Peace and the Twentieth Century Fund, (New York: Carnegie Endowment and Twentieth Century Fund, 1971), p. 18.
7. Ibid., pp. 18-19.
8. U.S. Office of the President, "Memorandum for the Honorable Dean Burch, Chairman of the Federal Communications Commission," January 23, 1970.
9. For example see Wilbur Schramm et al., The New Media: Memo to Educational Planners (Paris: UNESCO, 1967); Communication in the Space Age: The Use of Satellites by the Mass Media (Paris: UNESCO, 1968).
10. Alexander Gerschenkron, Economic Backwardness in Historical Perspective: A Book of Essays (New York: Praeger, 1962).
11. San Francisco Chronicle, June 5, 1974.
12. See Wall Street Journal, February 20, 1974.
13. See "Arabs Plan Domestic Satellite System," Aviation Week & Space Technology, September 30, 1974, p. 22.
14. San Francisco Chronicle, June 5, 1974. For further detail on the Indian system see P. P. Kale, R. L. Nickelson, and F. W. Sarles, Jr., "A Design for INSAT," in Bargellini, Communications Satellite Systems (Cambridge, Mass.: M.I.T. Press 1974); Vikram Sarabhai, "Television for Development," (Paper delivered at the Society for International

Development Conference, Delhi, November 14-17, 1969); Vikram Sarabhai et al., "INSAT—A National Satellite for Television and Telecommunication" (Paper delivered at Session 9 of the National Conference on Electronics, Bombay, March 24-28, 1970); E. V. Chitnis, K. S. Karnik, and B. S. Rao, "Satellite Television—A Formidable Tool for National Development," mimeographed, Indian Space Research Organization, 1971.

15. INTELSAT Agreement, Article IX, pp. 22-30, *passim*.

CHAPTER

8 COMMUNICATIONS TECHNOLOGY

INTELSAT AND SPACE AND COMMUNICATIONS TECHNOLOGY IN OTHER COUNTRIES

Many observers have complained that the United States wrongly exploited its monopoly in satellite and alunch technology in the mid-1960s by more or less forcing other countries to enter into the interim INTELSAT agreements on American terms:

> The United States was generous in wanting to share . . . the ability to communicate on a global scale. But the real treasures—the technological benefits to be gained from developing and building the system—were not offered and, in fact, were jealously guarded.[1]

What steps have been taken by INTELSAT and its members to assure the transfer and development of satellite-related technology to industrialized countries other than the United States?

Institutional Developments

To consolidate their bargaining power, the European countries formed the CETS (French initials for European Council on Satellite Telecommunications) group to negotiate with Comsat and the U.S. government as the interim agreements took shape in 1963 and 1964.[2] From the beginning, the European countries, as well as other industrial powers, such as Japan, Australia, and Canada, sought to maximize their share of INTELSAT contracting, knowing that only in this manner could their domestic space and telecommunications industries advance in the satellite area. Their struggle to do so, in retrospect, has been more fiercely waged and more successful than their efforts to thwart the U.S. policy favoring a single global commercial system.

Indeed, those countries saw the problem from the first in political terms, while the Americans had conviently decided to take a more narrowly commercial approach. The influential Causse Report, issued by a committee of the European space conference in late 1967, observed:

> The fundamental question is above all more political than economic in nature: do the European governments consider it necessary and desirable to devote the necessary resources to [space] programs in order to secure an independent long-term capability and not to remain tributaries of American facilities or dependent on global systems in which the United States has a preponderant influence?[3]

This realistic view of the political side of the technology was a curse as well as a blessing for Europe, however. There was no truly "European" space technology, but only an imperfectly blended brew of a number of national space policies. One unfortunate manifestation of this was the policy of *juste retour*, or pro-rata geographic distribution of contracting in participating countries, which was applied to ESRO and ELDO activities with a rigidity that greatly damaged the effectiveness of these two organizations:

> Comsat insisted that INTELSAT purchase only the best equipment available at the least expense. . . . Europe wanted contracts based roughly proportional to investment and ownership quotas in INTELSAT, regardless of cost. . . . The Europeans preferred the system of *juste retour* for INTELSAT.[4]

A careful and thoughtful study by a French author, Jacques Tassin, in 1970 criticized the proportional contracting idea endemic to European multinational undertakings:

> The notion of "*juste retour*" constitutes the alpha and omega of industrial policy [in European space technology] if not protectionism pure and simple, as for example in ELDO, where every country reserves the right to place by itself contracts for the organization.[5]

Tassin noted further:

> Although the attachment of member countries to [the principle of *juste retour*] is understandable, its application became so literal and petty [*mesquine*] that it did more harm than good to

> Europe. The "*juste retour*" has in effect dissipated [*fractionné*] the work to an excessive degree.[6]

Tassin recommended as an alternative the establishment of a fund to subsidize technology in countries which could not compete on a cost basis. They were to be awarded subsidized contract work in areas where failures or setbacks would have the smallest overall effect on the multinational project in progress.[7] Part of the failure Tassin lamented was of course the effort to develop an all-European launcher, an effort which was discontinued, as noted, after eleven successive launch failures. The ultimate frustration of the launcher program has been attributed to the insistence by ELDO members on contract participation strictly proportional to national financing quotas.

A high NASA official captured in the following words the basic early conflict in INTELSAT between the United States, anxious to plow forward with its monopoly of technology, and the Europeans, divided among themselves and fearing the unknown:

> Once the question of organizing a given space application is brought into an international forum . . . . a basic difference of view must be expected between the few space powers and the majority who are not space powers. A handful of countries will have relevant experience and knowledge. Depending upon the forum, as many as 120 nations will lack both. The practitioners and nonpractitioners will not have the same understanding of the space programs under discussion. . .
>
> In principle, we fear what we don't know. In the case of space, only a handful know; the great majority do not know. Thus, in a given international forum, fear and suspicion will outweigh confidence and assurance. And there will be some who are quite willing to exploit this situation.
>
> This means that the basic orientation of international forums is likely to be formalistic rather than operational, cautious where the state of the art may warrant action and enthusiastic where caution may be required. Above all, there will be a tendency to be regulatory and restrictive rather than facilitative. A space power prepared to proceed largely on its own and confident of producing positive benefits must, in other words, be prepared for a less positive reception than the technical prospects may seem to deserve.[8]

But perhaps the criticisms above are overly political, institutional, and psychological in approach. The simple fact is that for whatever reasons, Europe did not see fit to appropriate even a small fraction of the amount of money for space research and technology, including satellite communications and their launchers, that the United States was devoting to that area throughout the 1960s. The same NASA official offered this critique in 1967:

> It is clear that the European countries . . . have not yet taken the fundamental actions which alone can bring them to international bargaining tables as equal partners. They have not invested the time, money, and interest to put themselves in a basic position for takeoff in advanced applications . . . Europe is investing under $300 million a year in space technology relative to our $5 billion.[9]

For the entire decade of the 1960s, ELDO and ESRO expenditures were some $700 million, less than one-fifth of NASA's budget in 1969 alone.[10] Naturally, Europe's progress in space through ELDO and national efforts is redressing this balance during the 1970s.

To sum up consideration of the institutional aspects of European space programs and technology since INTELSAT began operation, it is worthwhile to cite Tassin once more, this time regarding the three possibilities he saw for European space technology vis-a-vis the United States in 1970. The first possibility, he said, was to acquire a "minimum autonomy" in space; the second was to "renounce that ambition and direct its efforts toward a complementarity with the American effort," which would entail "even greater dependence"; and the third and "most difficult" course would be to "create conditions of autonomy" and confront the United States "not as a client but as a partner."[11] The evidence since that time is mixed as to which course has been taken. Offered a share of the post-Apollo applications program by NASA, the European countries were unable to attain the funding or political unity to respond to the invitation in a meaningful way. On the other hand, the suggestions of some[12] that European space organizations unite organizationally to prevent dispersion of efforts and focus are about to bear fruit. In early 1975, as noted previously, ELDO and ESRO were close to a merger into a single European Space Agency,[13] the convention of which sets its goals as

> cordinating the common European space program and the national programs, and . . . integrating the latter progressively and as completely as possible into the common European space program, in particular as regards the development of application satellites.[14]

Europe, as noted above, still has a reasonably ambitious satellite program under preparation for the late 1970s. Furthermore, the French have been successful in inducing eight other European nations to join under ESRO's aegis in the development of the originally all-French L-3S launcher after the demise of the Europa-2 program.[15]

Industrial Developments

Parllel with the institutional and political initiatives described previously, the emerging European space technology has formed a number of effective lobbies and industrial interest groups that seem in retrospect to have been more effective than governmental bodies in promoting advances in European space and satellite communications technology. The first important such organization was EUROSPACE, formed in 1960 and comprising about 100 European industrial companies in the aerospace and electronic fields.[16] This organization was instrumental in the formation in 1972 of EUROSAT, a lobby of 55 firms from 9 European countries with an initial subscribed capital of about $1 million. The purpose of EUROSAT was essentially to stimulate European governments to study and adopt proposals for European regional satellite systems.[17]

In addition to these broad industrial lobby groups, individual consortia of two or more European aerospace firms have been formed, sometimes for long periods of time, to accomplish specific purposes. These transnational consortia have figured prominently in contracting activities both for INTELSAT and for all-European projects. For example, when Hughes Aircraft and Lockheed were assembling foreign participation in order to bid for the INTELSAT V program, each had its own list of European subcontractors formed largely along the lines of these preexisting transnational aerospace consortia.

Participation of Industrial Countries in INTELSAT Contracting

Within the INTELSAT framework, the main engine of transfer of satellite technology to countries other than the United States has of course been participation by those countries in INTELSAT contract work. This has consisted in the main of subcontracting by U. S. prime contractors building the INTELSAT I, II, III, and IV series, but has also included substantial research and development contracts to individual non-U. S. firms or groups of firms as well as operational tasks such as tracking, telemetry, and command of the INTELSAT system. In addition, there is a program whereby telecommunications experts from various countries may spend a year or more in visiting assignments at Comsat Labs.

Non-U. S. participation in the INTELSAT I series, begun before INTELSAT even existed, was nil. It was 2.3 percent in INTELSAT II and only 4.6 percent in INTELSAT III. The figure climbed to about 26 percent for the first four satellites of the INTELSAT IV series, for which Hughes Aircraft subcontracted $19.4 million abroad. The final four satellites in this series, however, were acquired under circumstances in which foreign contracting was closer to 10 percent.[18] A subsidiary issue has involved access of firms contracting with INTELSAT to data and patents arising from those contracts. Some such firms, particularly European ones, have withdrawn after complaining that restrictions on their rights to patents and data have been too severe.

Despite pressures from industrial countries outside the United States, both the interim and permanent INTELSAT agreements have avoided requirements which would place the interest of _juste retour_ above that of lowest cost, highest quality, and best delivery schedule. Article X of the Interim Agreement of 1965 provided:

> . . . the [Interim] Committee and the [Communications Satellite] Corporation shall be guided by the need to design, develop and procure the best equipment and services at the best price for the most efficient conduct and operations of the space segment. When proposals or tenders are determined to be comparable in terms of quality, . . . price and timely performance, [they] shall also seek to ensure that contracts are so distributed that equipment is designed, developed and procured in the States whose Governments are Parties to this Agreement in approximate proportion to the respective quotas of their corresponding Signatories to the Special Agreement.[19]

Despite similar pressure, the permanent agreement contained even vaguer language and by no means enshrined the _juste retour_ principle as it has existed, for example, in ill-fated European organizations such as ELDO and The European Atomic Energy Community (EURATOM):

> . . . procurement of goods and services required by INTELSAT shall be effected by the award of contracts, based on responses to open international invitations to tender, to bidders offering the best combination of quality, price, and the most favorable delivery time. . . . If there is more than one bid offering such a combination, the contract shall be awarded so as to stimulate in the interests of INTELSAT, world-wide competition.[20]

Interests of Developing Countries in INTELSAT Contracting

The European argument for more foreign procurement in INTELSAT contracting reduces to an infant industry argument based not on tariff protection but on preferential contracting procedures which ignore cost minimization in favor of the _juste retour_ principle. In terms of their own self-interest, the Europeans found the short-term costs of higher INTELSAT rates due to higher INTELSAT costs less than the political and long-term economic losses of preclusion from extensive participation in INTELSAT contracting and the consequent damaging effects on their own domestic space industries.

Thus, the interests of the United States on the one hand, and of the developing countries in INTELSAT on the other, the developing countries, with little to hope for in the way of contracting or technology, tended to form a common front against European desires for cost-increasing internationalization of procurement participation. As early as 1968, for example, when foreign contracting was still minimal, Argentina suggested to the ICSC that member developing countries excluded from subcontracting in the INTELSAT IV program be compensated for their effective subsidization of European industry by means such as training programs, scholarships, and research grants.[21] And after the construction of the first four INTELSAT IV satellites, Comsat complained that non-U.S. participation had resulted in at least a 5 percent increase in overall costs.[22]

The gap between space technology in the United States and other industrialized countries has in fact lessened to some extent, and might have lessened even more had preferential INTELSAT procurement policies been followed. A price would have been paid, of course, in terms of the sacrifice of static present gains from lower INTELSAT rates for the dynamic, less easily quantifiable benefits of accelerated technological progress over a period of time. In any event, the United States was able to preempt much of the early technology and contract participation in the satellite area not solely because of alleged machinations by Comsat and the U.S. government in the first INTELSAT negotiations, but because of the tremendous amount of money the United States, both publicly and privately, had already devoted to that area. As noted, Western Europe, with human and material resources comparable to those of the United States, chose to devote only a small fraction of tis resources to those ends, and a technological gap which still persists, though in attenuated form, is the result. Indeed, the U.S. taxpayer subsidized international satellite communications from the first, to the extent American military satellites, launchers, and other space and communications activities had preceded INTELSAT and provided externalities to INTELSAT not reflected fully in its costs and prices. Europe, as noted, chose, perhaps wisely, not to be a passive beneficiary of that subsidy at the cost of developing its own satellite industry and aerospace technology, although strictly economic considerations might have dictated the opposite course.

CABLES AND SATELLITES

While one tends to think of cable technology as having preceded satellite technology substantially, it is noteworthy that the first voice-quality cable was not laid across the Atlantic until 1956, only two years before the first experimental communications satellite was launched.[23] Before that time, only telegraph cables, some of them dating from the late 1850s and of very low capacity and dependability, had been available. The exponential capacity increases in both cables and satellites during the 1960s and 1970s have contributed to a very confused and convoluted series of issues regarding their relative costs per channel. No effort will be made at this point to delve into those issues, because this study has been concerned only with the space segment of international satellite communications rather than with the earth stations or terrestrial distribution costs to the ultimate consumer. It bears mention, however, that most studies in this area[24] conclude that there is a threshold distance beyond which satellites, which are not cost-sensitive to distance, have lower per-channel costs than cables, which are. The threshold distance is generally on the order of 600 to 2, 000 miles, which is much less than links generally provided by INTELSAT.

A particular country's preference for cables or satellites will depend on much more than a priori comparative costs. The market and governmental structure relating to satellite communications is also a critical factor.[25] In the United States, the main cable power, AT&T, now that it has sold its entire interest in Comsat, has no rate base advantage* to gain from leasing satellite circuits from Comsat but derives a substantial rate base advantage by increasing its investment in new cable systems. Thus, the authorization of the TAT-5 and TAT-6 cables by the FCC was strongly opposed by Comsat and strongly urged by AT&T. In most countries, by contrast, both cable and satellite activities are subsumed under a single ministry of telecommunications, which often does not have the manpower or at any rate the economic incentive to clearly define its economic interests, to which in any event it might judge other interests, such as political and technological advantage, superior. Many of the larger European countries, such as Great Britain and France, have substantial investments in existing cable systems. For example, before certain African countries built earth stations, calls between them had to be routed through one or even two European capitals, following existing cable (and colonial) links, with consequent transiting fees to the European countries involved.

---

*The advantage referred to is that of the Averch-Johnson effect, referenced earlier in this study, whereby rate-base regulation confers a private incentive to capital expenditure greater than that which is socially optimal.

From an overall welfare standpoint, economic theory would require that decisions as to whether cable or satellite facilities were to be added to fill additional international demand at any given moment be made solely on the basis of the future marginal costs such decisions would incur, with no attention whatever to sunk or past costs.[26] Yet structurally, the pattern of ownership and administration of the international telecommunications system is so diverse and fractionated, and the area of telecommunications itself so fraught with noneconomic implications, that there is little hope such a simple criterion would ever prevail. For example, in one important case involving authorization of new capacity, the FCC decided for a time on the future equal sharing of traffic by satellite and cable facilities, apparently as a purely political compromise between the parties involved. Rate-base considerations, as noted, distort private from public costs for regulated concerns along the lines elaborated by H. Averch and Leland L. Johnson in the early 1960s. And administratively, the operating agencies which are signatories to the INTELSAT Operating Agreement range from all-satellite concerns, such as Comsat in the U. S. and Telespazio in Italy, to telecommunications ministries of smaller countries, and to those of European countries with substantial cable interests. Institutionally and financially, there is no imposition of a global welfare criterion by which a least-cost decision regarding the relative economic merits of cables and satellites could be made. Patterns of cross-subsidization in some countries further complicate the picture, as telecommunications revenues from the often lucrative overseas cable and satellite traffic are used in many countries to subsidize chronic operating deficits in domestic postal and telephone services.

Finally, as Joseph N. Pelton has noted, the omission of earth stations from INTELSAT ownership and finances "served to undercut the efficiency of the system,"[27] in that tradeoffs between the interdependent costs of launches, satellites, and earth stations cannot be made from an overall standpoint. The same may be true for cables, since cableheads are owned by individual countries.[28] From the standpoint of economic theory, however, vertical integration of the international satellite system is necessary only if price fails to equal marginal cost in any of the presently unintegrated subcomponents, such as the space segment, the earth stations, and the terrestrial infrastructure. If price did equal marginal cost in each subcomponent, there would presumably be no efficiency gain from integration. Institutionally, however, satellite communications has in INTELSAT the advantage of an international forum in which global welfare criteria can at least be discussed, if not always implemented. No such organization exists for the world's undersea cable system, which is characterized by a multiplicity of bilateral relations between different users.

NOTES

1. Judith T. Kildow, INTELSAT: Policy-Maker's Dilemma (Lexington, Mass.: Lexington Books, 1973), p. 16.

2. See Jonathan F. Galloway, The Politics and Technology of Satellite Communication (Lexington, Mass.: Lexington Books, 1972), pp. 94-95.

3. European Space Conference, Report of the Advisory Committee on Programmes, December 1967, CSE/SSP(67)5, final version, quoted in Jacques Tassin, Vers L'Europe Spatiale (Paris: Denoel, 1970), p. 176. Translation by the author, for this and all further quotes from this work.

4. Kildow, op. cit., pp. 52-53.

5. Tassin, op. cit., p. 17.

6. Ibid., p. 145.

7. Ibid., pp. 147-49.

8. Arnold Frutkin, "Patterns of International Space Applications and Their Extension" (Speech given at the American Astronautical Society, 13th Annual Meeting on Commerical Utilization of Space, Dallas, May 2, 1967), quoted in Kildow, op. cit., pp. 24-25.

9. Frutkin, ibid., quoted in Kildow, op. cit., p. 28.

10. Tassin, op. cit., p. 145.

11. Ibid., p. 19.

12. For example see ibid., p. 194.

13. For the background of the merger, see "Europe Delays Shift to Single Space Agency," Aviation Week & Space Technology, April 1, 1974, p. 21; and Communications Satellite Corporation, Geneva Office Handbook, October 1973, p. 273.

14. "Europe Delays Shift to Single Space Agency," Aviation Week & Space Technology, April 1, 1974, p. 21.

15. See Robert R. Ropelewski, "Demise of Europa-2 Clears Way for L-3S," Aviation Week & Space Technology, May 7, 1973, p. 20, and Ropelewski "Europe L-3S Launcher Start Set for 1974," Aviation Week & Space Technology, October 1, 1973, pp. 34-36.

16. Communications Satellite Corporation, op. cit., October 1972, p. 148.

17. Ibid., pp. 146-47.

18. Kildow, op. cit., pp. 53-54.

19. INTELSAT Interim Agreement, Article X.

20. INTELSAT Agreement, Article XII(a)-(b), p. 42.

21. Pelton, Global Communications Satellite Policy: INTELSAT, Politics and Functionalism (Mt. Airy, Md.: Lomond Books, 1974), p. 47.

22. Kildow, op. cit., p. 53.

23. Pelton, op. cit., p. 47.

24. See, for example, U.S., Office of the President, Office of Telecommunications Policy, "International Facilities Study," May 1971.

25. See Pelton, op. cit., p. 47.

26. See Alfred E. Kahn, *The Economics of Regulation, vol. 1, Principles* (New York: John Wiley, 1970) p. 73.

27. Pelton, op. cit., p. 59.

28. Ibid.

PART IV

A SUMMING UP

CHAPTER

9 INTELSAT AND WELFARE

The mention of global welfare criteria in the preceding chapter is a convenient point of departure for analyzing INTELSAT from a welfare standpoint. Has INTELSAT had particular criteria of welfare as tacit or explicit goals, are they quantifiable, and how successful has INTELSAT been in reaching them? Would pursuit of different criteria have increased welfare in any meaningful sense? For the moment we take "welfare" in the everyday sense rather than in its narrower economic usage.

Not all of these questions can be answered, or even guessed at intelligently. Perhaps the best place to begin is to suggest a number of welfare goals INTELSAT might be assumed to have pursued, in order to focus on which of them might be the most reasonable criteria from which to judge INTELSAT's performance.

POSSIBLE INTELSAT GOALS AND WELFARE CRITERIA

In this chapter, alternative goals will be suggested and discussed briefly. Strictly speaking, it makes little sense to speak of "INTELSAT's" goals pure and simple, so one must tacitly assume some sort of social welfare function or political process by means of which the goals and strategies of individual INTELSAT members are translated into actions of the ICSC or the Board of Governors.

Cost Minimization

The conclusion that INTELSAT has been a cost-minimizer during its first decade of operation would be a quite attractive one if it were not for some evidence of chronic excess capacity during that time. As has been argued elsewhere,[1] however, even observed excess capacity must to some extent be interpreted as necessary backup in the case of failure of in-orbit major path satellites. In that event it is not excess at all, but part of the high-dependability package which the presumably risk-averse customer purchases from INTELSAT. As noted, satellites are vulnerable to occasional "catastrophic" or total failures in orbit and cannot be repaired, at least by physical contact. In addition, many INTELSAT members have a financial and administrative incentive to view cables and satellites as part of a single global communications system, whereas INTELSAT financial records do not reflect this integrated incentive for reciprocal backup of spare capacity by the two modes.

Several possible causes of observed excess capacity, however, cannot properly be reconciled with cost minimization as a goal of INTELSAT. Examples include (1) overcapitalization along the lines of the Averch-Johnson thesis, with the incentive to overcapitalization coming from those members of INTELSAT (Comsat primary among them) which are subject to rate-base profit constraints from their own national administrations; (2) overestimation of future traffic due in part to delays in completion of new earth stations; (3) a desire to duplicate the latest INTELSAT satellites in all ocean regions as soon as possible, sometimes without full regard for varying regional traffic patterns; (4) less than maximally efficient usage of nominal capacity due to small traffic links, effectively reducing the capacity available; and (5) a desire to build quickly and ahead of demand in order to preempt construction and launch of possible rival systems, whether domestic or regional.

While it is undeniable that each of these latter causes has contributed in some measure to observed overcapacity during INTELSAT's first ten years, most of the deviation between traffic and capacity apparent in INTELSAT (see Table 2.1) can be confidently attributed to backup capacity for use during satellite (or cable) failures. It is common telecommunications engineering practice to furnish twice as much capacity as will be required (abstracting from peak-loading problems) over a given period. To the extent, then, that this has been the predominant cause of its excess capacity, INTELSAT's investment and operating behavior since it began operation can be quite easily interpreted as consistent with the goal of minimization of per-channel costs to users.

Rate of Return

As shown in the Appendix, the rate of return chosen by INTELSAT is essentially irrelevant to the extent ownership equals use on the part of INTELSAT members. It has been important, however, as a kind of barometer, particularly in the early days of INTELSAT when there was doubt as to whether the new technology could be commercially successful.

INTELSAT has never acted as though its rate of return were in and of itself an important issue, although particular members with imbalances of use over investment (or investment over use) might have wished for a a rate of return lower (higher) than 14 percent. It is important to distinguish the financial incentives of INTELSAT signatories as members of a cooperative in which their ownership and use are equal from their incentives as wholesalers or retailers of satellite circuits to their own domestic markets. In the former capacity, as shown in the Appendix, the members gain nothing from increasing their rate of return as a camouflage for supernormal profits, assuming investment is equal to usage. But in the latter capacity, at least if they are profit-oriented firms as opposed to public agencies, INTELSAT members have the usual monopolist's incentive to maximize profits by increasing price to the point where marginal revenue equals marginal cost. In this case, however, the cost derives from the signatory's participation as an INTELSAT owner-user, while the revenue is generated by the resale of INTELSAT's circuits to a domestic market in a transaction which is outside the financial and administrative framework of INTELSAT. Thus, ceteris paribus, profits for INTELSAT signatories increase as INTELSAT's investment and operating costs go down, thereby contributing to the cost minimization goal noted above.

Traffic Maximization

One could suppose that psychic and economic benefits would be derived from attracting the greatest possible amount of traffic, measured in units of utilization, to the INTELSAT system. INTELSAT, however, has not pursued this as a goal. In the first place, as noted in Chapter 8, most INTELSAT operating agencies also represent cable traffic financially or administratively and have no particular incentive to increase satellite traffic at the expense of cable traffic. Furthermore, it has been noted that satellite traffic is probably quite price-inelastic and would therefore not be particularly amenable to substantial increases by price reductions. As has been seen, the exception is the case of whole or half transponder leases directly for domestic usage with no intermediaries.

Capacity Maximization

INTELSAT has had excess capacity as a matter of course throughout most of its existence, as noted. Some of the reasons noted in this regard, such as redundancy to provide continuity in the case of satellite or cable failure, reflect strongly held views on the part of the communications establishment and must therefore be reflected in any accurate welfare criterion. Other reasons, such as rate-based maximization on the part of INTELSAT members subject to domestic rate-base regulation (primarily Comsat), reflect special interests not ascribable to INTELSAT as a whole. Some reasons have been ascribed to stochastic or accidental factors such as overestimation of demand and do not enter into any conscious welfare criterion. But INTELSAT's behavior certainly does lend at least some support to the view that capacity maximization has been pursued as an independent goal.

Maximization of Technological Progress

Economic theory does not hold technical sophistication as a goal in and of itself, but rather prescribes an optimum level of technology for a given set of factor prices, consumer demand, and so forth. It could very well be that a particular technology (a common example is the extensive use of tractors in a developing country with a poorly trained and largely unemployed labor force) is inappropriately exotic in given situations. Yet a case has been made elsewhere[2] that, at least through the INTELSAT IV series, INTELSAT's level of technology has not been inappropriately high, and that in fact the form in which its technology has advanced—primarily the extension of scale frontiers to realize even greater scale economies—was the main constraint on lower average system costs. Only now is INTELSAT at a crossroads, it would appear, where a broad range of technological choices is present, many of which involve a good deal more in the way of technological progress than further rolling back of scale frontiers.[3] In this context, INTELSAT's choice to go ahead with the INTELSAT IV-A program for the time being rather than with the INTELSAT V is clearly a decision not to adopt a technology needlessly exotic from the standpoint of cost minimization.

Prestige and Institutional Enhancement

John Kenneth Galbraith and others have suggested a kind of perquisites-and-prestige maximization goal as a credible substitute for profit maximization in some industries: the "technostructure" desires to secure its "autonomy."[4] Is there evidence that INTELSAT member states, or more particularly, individual diplomats, scientists, officials, and experts closely associated with INTELSAT, have sought to attain private goals of this nature, perhaps at the expense of cost minimization? One would be hard-pressed to find evidence to support this view. INTELSAT has only a small permanent staff, although it will expand along the lines of the permanent agreements. In its early years, INTELSAT's work was done by Comsat employees charging part of their time to INTELSAT. Board of Governors members currently pay their way to meetings with funds of their own administrations, not those of INTELSAT. INTELSAT kept an extremely low profile during its development, and some members finally complained that it was not even listed in the District of Columbia telephone directory. Physically, INTELSAT used to consist of half of the third floor of Comsat headquarters in Washington, although it has now moved to its own quarters in the building next door. One can confidently reject the suggestion that pursuance of its own prestige or institutional enhancement, or that of its representatives and experts, has played a significant role in INTELSAT's development.

World Peace and Understanding

Although the INTELSAT Agreement contains a reference to INTELSAT's contributing to "world peace and understanding,"[5] there is little evidence that INTELSAT's behavior with respect to that goal has been pursued at the expense of cost minimization or other financial criteria. As noted previously, INTELSAT's policy on free usage of the space segment has been somewhat restrictive. Once, for example, a U.N. request for free usage of channels for peacekeeping operations in the Middle East was met with delay because of a feeling among most INTELSAT members that the United Nations should pay for the circuits. Finally, INTELSAT's activities as noted in Chapter 7, have been oriented much more toward the conventional telecommunications needs of industrialized countries than toward special needs of the less developed world.

Facilitation and Maximization of Participation by Different Countries

Resolution 1721(XVI) of the General Assembly of the United Nations, quoted in the Preamble to the INTELSAT Agreement, calls for satellite communications to be available "as soon as practicable on a global and nondiscriminatory basis" to "the nations of the world." The U.S. Communications Satellite Act speaks similarly of the rapid and broad proliferation of satellite communications. Has the speed or extent of the proliferation of satellite communications been at the expense of any criteria above, particularly that of cost minimization?

This question must be posed from a number of different standpoints. Consider first the speed with which satellites came into operation after they became technically feasible. Arguments have already been noted to the effect that the United States rushed into INTELSAT with its own satellite, Early Bird, and strove to preempt the institutional power and technical superiority of satellite communications by bringing off fait accompli in which other nations would be obliged to acquiesce. Furthermore, some of the excess capacity of the kind INTELSAT has experienced might reflect this kind of haste, as has been suggested above.

The issue of diffusion of INTELSAT usage cannot be treated with reference to the space segment alone. It must be posed with reference to each INTELSAT member that built one or more earth stations to gain access to the INTELSAT space segment. How many of these countries build earth stations which from a purely economic standpoint were premature and too costly? Did factors of prestige, international peer pressure, and the allure of new technology play a factor on the demand side? The answer can only be obtained by examining each INTELSAT member country to see if costs for international telecommunications rose or fell as the result of access to the INTELSAT space segment, in comparison to costs of other possible strategies (establish a link to a new cable, for example). There are indications that at least a few developing countries committed themselves to INTELSAT membership and use with levels of international traffic so low that their earth stations and INTELSAT capital contributions could not be supported by domestic traffic revenues. Pelton, for example, states that 15 circuits of traffic comprise the usual threshold for earth station financial viability and, noting that 34 of INTELSAT's 83 members in March 1972, had investment quotas of 0.05 percent or below, ascribes their membership to "status motivation" rather than "sound financial investment."[6] On its part, however, INTELSAT has exerted no particular pressure to maximize its membership, although it has welcomed all new members as they came along. Comsat, for its own purposes, has been anxious to maximize INTELSAT membership and to provide consulting services on earth station viability and operation to prospective member countries. Yet most countries with any appreciable degree of traffic are now INTELSAT members, with the exception of socialist countries. These would probably constitute the last frontier for INTELSAT membership, but no serious overtures are apparent.

THE BASIC CHOICE: POLITICS OR BUSINESS AS USUAL

As indicated above, cost minimization, even in the face of apparent excess capacity, has been the major detectable goal of INTELSAT, and that capacity maximization and to a lesser extent the facilitation and maximization of participation in INTELSAT are the only other goals that appear to have figured prominently and to the partial detriment of cost minimization. Chapters 6-8, which dealt in part with separate systems, technological development and international procurement, and the interests of developing countries, shed some light on the extent of these subsidiary goals. Many goals, of course, have not conflicted with cost minimization; examples include some degree of traffic maximization and technological progress.

A fundamental choice with which INTELSAT has been faced is, in retrospect, that of engaging in a conventionally "commercial," cost-minimizing activity as opposed to other activities that must for brevity if not for accuracy be termed "political." Political goals might have included free channels for development-oriented programs, subsidization of space industries in European countries, extensive politicization of INTELSAT through the permanent agreements, an earth station and INTELSAT membership for every country regardless of traffic level, and so on. By and large, INTELSAT's essentially commercial orientation has not been sacrificed for political goals of this nature, and despite the large voice given to small users in the permanent agreements, it does not appear that the future will bring any real change to INTELSAT's basic character. Its "prime objective" is still the provision of capacity for "international public telecommunications on a commercial basis."[7]

The primacy of commercial over political goals has struck different observers in different ways, and a review here of conclusions reached by three writers who have examined INTELSAT at some length will serve to cover the range of opinion.

Brenda Maddox, in an almost wholly negative assessment of INTELSAT written in 1972, lamented that "INTELSAT might have led the world in the planning for the use of satellites for education and for the exchange of television programs."[8] Further, she noted:

> INTELSAT is overstructured, overdefined, and underpowered. The scope of each layer of the organization has been so rigidly limited that INTELSAT will find it difficult to make any drastic changes of direction or to do much more than to sell telecommunications to those who have the money to pay.
>
> INTELSAT . . . is a badly bungled international opportunity. It was a chance to go beyond Babel, to use the new technology of satellites to creat an organization in which governments not particularly friendly toward one another might have worked together. . . . INTELSAT is the triumph of the past over the new technology, a narrow commerical organization with an accountant's mentality.
>
> In the unlikely event that there is ever a child who asks, "What is INTELSAT, Daddy?" the answer can only be, "It is the big telephone company in the sky."[9]

Judith Kildow's treatment is that of a political scientist rather than a journalist, but her assessment of INTELSAT is only slightly more charitable than Maddox's and focuses likewise on the conflict between political and commercial goals. In general, according to Kildow, there were "two intertwining central issues" in the establishment of INTELSAT, the first "centered on the U.S. objective of a single global communications system, as opposed to a series of separate domestic and regional systems." The second issue

> was over a decision-making process based primarily on a commercial philosophy (advocated by the Americans), as opposed to one based on political participation (advocated by the Europeans).[10]

And Europeans, it appears, are not persuaded by the "very logical" but "commercially-oriented American arguments" but instead consider "long-term political and local economic considerations" paramount.[11] Finally, businessmen and engineers generally triumph over politicians when a new technology is involved:

> The close affiliation of political and economic factors [in U.S. foreign policy generally] has made consideration of both inseparable for most policy decisions. However, there are internal contradictions between the two perspectives that can lead either to indecision or inconsistencies in policies. In technology-related areas of foreign policy, this problem can become especially acute. When economic considerations dictate one course of action and political considerations another, the close relation of economic interests to technical developments often leads to weighting decisions in favor of commerical-economic objectives and has the consequence of relying on advice—in matters of foreign policy—from corporate executives and technicians.
>
> The question is whether the political costs of alienating others and dividing the international community are compensated by foreign appreciation of the technological results, or whether the economic costs of permitting political participation would have been too great to bear.
>
> There is really no way to measure either of these costs or to resolve the problem of which approach is best. In the case of INTELSAT, one can only evaluate the effects of pursuing a philosophy based on economic efficiency. . . . Perhaps the system was developed and established too soon, before a social and political framework could deal with it properly.[12]

A detailed study by Pelton also approaches INTELSAT from the standpoint of political theory, in this case as a critique of the theory of functionalism propounded by David Mitrany and refined by Ernst Haas.[13] Yet it differs strongly from both the Maddox and the Kildow treatment in reaching generally favorable conclusions regarding INTELSAT's goals and orientation while agreeing that INTELSAT has been more oriented toward a "quasi-commercial" approach than toward "welfare," "integrative" or "political" goals. While Haas and Mitrany argued that the "potential triumph of welfare over political values,"[14] and that "functionally specific international programs, if organized separately from defense, maximize both welfare and integration,"[15] Pelton, examining INTELSAT as a test case for functionalism, found that that organization showed "a wide-ranging ability to contribute, be neutral, or undermine welfare and integration, with the normal experience being that of neutral activities."[16] Having weakened the opposition posed by functionalists between "political" values on the one hand and "integrative" and "welfare" values on the other, Pelton opposes all of these to the quasi-commercial nature of INTELSAT and finds the latter criterion more appropriate for "international Functionalist organizations" in general:

> The INTELSAT experience would suggest that . . . new international scientific organizations may develop more quickly and efficiently if organized on a "quasi-commercial" rather than on "welfare" or "integrative" bases. This does not mean that direct governmental participation is to be or should be excluded, but rather decisions would be made in terms of financial or technical objectives regardless of political goals or long-term ideals. INTELSAT was formed less than ten years ago and now dominates international communications even in a competitive arena. In contrast, INTERSPUTNIK, EURATOM, and ELDO, to name only a few examples, indicate what the alternative might have been in INTELSAT had been based on political or broad policy goals.[17]

Pelton concluded:

> . . . the INTELSAT Organization should prove to be a useful prototype not only for new international scientific and technological ventures, but for the entire pattern of future global economic relationships.[18]

These three studies agree, if on no other point, at least in the conclusion that INTELSAT has been essentially commercial rather than political (integrative, welfare-oriented) in nature. Yet this dichotomy does not exhaust all possibilities. The commercial criterion of efficiency, strictly speaking, concerns only the maximization of profit, which

FIGURE 9.1

INTELSAT Costs, Output, and Demand

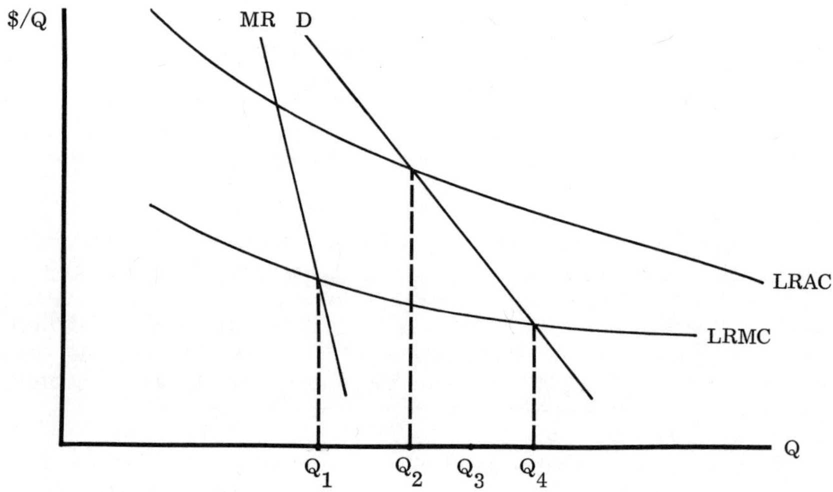

reflects only the interests of stockholders (INTELSAT investors). The ultimate economic criterion from which INTELSAT's performance should ideally be judged, however, is the maximization of the present value of net consumer surplus or benefits, assuming income distribution among consumers is considered satisfactory.

The best way to determine which of these three criteria has been most consistently reflected in INTELSAT's performance is to consider the output selection mechanism INTELSAT may be presumed to have followed. This question was raised above in regard to whether INTELSAT followed a policy of traffic and capacity maximization. Figure 9. 1 will provide a diagrammatic basis for further discussion.

The graph in Figure 9.1 indicates the declining long-run average costs determined in Chapter 2, the fact that marginal costs are less than average costs over the entire range of output, a steep (inelastic) demand schedule, and the associated marginal revenue curve. Various output levels are indicated on the horizontal axis which correspond to various of the welfare criteria discussed above. If, for example, INTELSAT members acted as unregulated, profit-maximizing monopolists in retailing INTELSAT circuits to their own domestic markets,* they would

---

*It should once more be stressed that transactions between INTELSAT signatories and domestic customers do not fall within INTELSAT's financial or administrative framework. In the discussion which follows,

they would select output Q_1, corresponding to the equality of marginal cost and marginal revenue, at which single-price profit maximization is attained. This would certainly be an extreme example of the commercial criterion, in which the interests of producers are paramount and those of consumers are ignored. Q_3, by contrast, is the output corresponding to the intersection of demand with long-run marginal cost. At that output level, the economic criterion of consumer welfare maximization in a Pareto-optimal sense may be said to apply, since price is set at long-run marginal cost. But it also entails an operating deficit because of declining average costs. At output level Q_2, price is equal to long-run average cost, so INTELSAT's costs (including return to capital) are just covered. This reflects a commercial welfare criterion, one similar to that allowed monopolists under conventional rate-base regulation, but tempered by the regulatory constraint of normal return and therefore to some extent economic as well.

The evidence presented in this study is consistent with the hypothesis that INTELSAT has produced at least up to the point that would correspond to Q_2 in Figure 9.1, at which price is equal to long-run average cost. This is implicit in INTELSAT's revenue requirements approach to price-setting. Furthermore, there is evidence that INTELSAT has gone beyond that point toward an output such as Q_4, in that it has begun to market spare capacity at a price below average cost on the basis of price discrimination favoring low-elasticity purchasers.

As has been noted, however, INTELSAT follows a strict policy of cost recovery, and it would be politically impossible for it to produce at level Q_3 because of the enormous subsidy that would be required to recover its costs. Thus INTELSAT, while appearing to operate on the quasi-commercial criterion of cost recovery plus a rate of return, might also be presumed to operate with respect to an economic criterion subject to the constraint of full cost recovery. Evidence for the latter position is its attempt to market excess capacity by discriminatory pricing, which is consistent with welfare-maximizing behavior. (Persistence of excess capacity is evidence that INTELSAT's behavior is still not wholly consistent with either an economic or a commercial criterion of welfare maximization.) In fact, the tentative conclusion is justified that, given the institutional, financial, and political constraints under which it operates, INTELSAT has performed rather well

---

INTELSAT is conceptualized as the source of supply, to which INTELSAT members happen to belong as a kind of cooperative. Thus, as noted, they all have an incentive for cost minimization of INTELSAT circuits. In retailing to their domestic market, they have an incentive for profit maximization, at least if they behave as private firms. The demand spoken of should be taken as the aggregate domestic demand for satellite communications to which INTELSAT members respond.

by an economic criterion. This conclusion is reached, it should be noted, on the basis of the output level it has selected. The evidence that INTELSAT has attempted to minimize costs, by contrast, is consistent with either a commercial or an economic criterion and therefore cannot be used to distinguish between the two.

In the political conception of INTELSAT urged by Maddox, Kildow, and most European countries, the organization should and should have served a variety of needs beyond the simple provision of international public telecommunications at minimum cost. For example, it should provide free circuits to the United Nations, subsidize the European space and telecommunications industry, support financially and institutionally the broadest possible expression of views of all its members, grant decision-making power on bases other than investment and traffic, abandon economies of scale to more expensive regional and domestic systems, and so on. In the commercial notion of INTELSAT espoused by Pelton, the United States, and most developing countries, INTELSAT should provide conventional and specialized communications at the lowest possible price, and to the extent its diminishing costs allow it to compete favorably with other modes. In the economic concept outlined here, the interests of consumers are to be maximized.

Once more, the economist's job is not to pronounce judgments on the relative merits of economic as opposed to political goals in satellite communications or in any other area. It is to welfare economics, in fact, that social science is indebted for the proposition that interpersonal welfare schedules are incomparable and therefore not amenable to aggregation into a collective social welfare function. There is, however, an essential incompatibility of political and economic goals in a functioning organization such as INTELSAT that must cover costs with traffic revenues.

INTELSAT is a wholly novel international organization. Instead of conducting studies, making declarations, or performing services for which is it not reimbursed, it owns and operates a revenue-bearing enterprise that has over a ten-year period enjoyed exponentially decreasing unit costs, exponentially increasing usage, several plateaus of technological advancement, and a cumulative recovery of all capital costs plus a high rate of return. This is an unprecedented accomplishment in the field of international organizations and should not be minimized by protestations to the effect that Comsat's early actions were those of brusque American businessmen, or that any number of desirable social goals were not accomplished as well. If the economics discipline has a message to offer, it is that INTELSAT has functioned well precisely because it has functioned on an economic basis, and that it would have functioned less well if it had simultaneously been required to fulfil noncommercial purposes without subsidy. Other goals, many of which are of surpassing importance, (for example, literacy and medical training by satellite television), can best be met either by entirely separate systems financed on a noncommercial basis, or, perhaps even more desirably, by a system under INTELSAT aegis, financed separately from public international traffic but benefiting from INTELSAT's past experience.

AREAS FOR FUTURE INQUIRY

The results of the present study serve to point the way for future contributions to the economic study of satellite communications. All of Part III, for example, dealt with an exposition of areas of interest, such as resource and development implications, technological progress in industrial countries other than the United States, and separate systems and the monopoly issue, each of which invite more extended examination from an economist's standpoint than could be devoted to it in the present study. Other areas of this study also raised questions for further inquiry with some promise of interest and success. For example, it was noted that careful studies of demand patterns for satellite communications in individual countries would provide a more reliable basis on which to draw conclusions regarding any policies of limited price discrimination which INTELSAT might choose to incorporate into its tariff.

Three final areas of future investigation might be mentioned. First, the question of cross-subsidization of INTELSAT's services was not dealt with in detail because of the limited range of conclusions regarding INTELSAT's underlying costs. If a method were devised more accurately to allocate INTELSAT's costs to individual services, rather than to deal with its output as an essentially homogeneous variable, patterns of cross-subsidization might be detected to a greater extent than they were in this study, and conclusions of an economic and political nature might be drawn. Second, as noted in Chapter 1, the focus of the present study is necessarily constrained to INTELSAT itself. Perhaps further analysis could use the present conclusions as a means of expanding the scope of study and applying broader welfare criteria in the conclusions reached.

Finally, it should be stressed that better and more complete data regarding INTELSAT's costs and technology are required for more powerful statistical methods to be applied. The passage of time will be a major ally in increasing the degree of freedom available for future statistical research. More years of INTELSAT operation will have elapsed, increasing the time series data available; furthermore, more satellite systems, such as the domestic U. S. systems now entering operation, will yield data to provide a cross-section. With such data, broader and more definitive conclusions will be possible.

NOTES

1. See Marcellus S. Snow, "Investment Cost Minimization for Communications Satellite Capacity: Refinement and Application of the Chenery-Manne-Srinivasan Model," *Bell Journal of Economics* 6, no. 2 (August 1975).

2. Ibid. Unit costs which decline up to full capacity, as established in the present study, also support this viewpoint.

3. See J. J. Knopow, "Next Generation Communications Satellites," and D. C. MacLellan, "Anticipated Developments in Communications Satellite Technology," in *Communications Satellite Systems*, ed. P. L. Bargellini (Cambridge, Mass.: M.I.T. Press, 1974).

4. John Kenneth Galbraith, *The New Industrial State* (Boston: Houghton-Mifflin, 1967), p. 171.

5. INTELSAT Agreement, preamble, p. 4.

6. Joseph N. Pelton, *Global Communications Policy: INTELSAT, Politics and Functionalism* (Mt. Airy, Md.: Lomond Books, 1974), p. 136.

7. INTELSAT Agreement, Article III(a), p. 9.

8. Brenda Maddox, *Beyond Babel: New Directions in Communications* (New York: Simon and Schuster, 1972), p. 82.

9. Ibid., pp. 110-11, 114.

10. Judith T. Kildow, *INTELSAT: Policy-Maker's Dilemma* (Lexington, Mass.: Lexington Books, 1973), p. 41.

11. Ibid., p. 65.

12. Ibid., pp. 85-87.

13. See Pelton, op. cit., pp. 9-16.

14. Ibid., p. 13.

15. Ibid., p. 155.

16. Ibid.

17. Ibid., p. 158; the "welfare" values spoken of here might correspond most closely to what an economist would refer to as "psychic" utility, that is, satisfaction derived from sources other than goods or services.

18. Ibid., p. 161.

APPENDIX: MODEL OF INTELSAT INVESTMENT AND USE

NOTATION

All quantities are expressed with respect to a given period of time:

D = total accounting depreciation
R = total return paid on capital by INTELSAT
O = total operating expenses
K = total capital contributions by signatories
N = total number of units of utilization sold (ex ante and ex post)
P = space segment utilization charge per unit
q_i = ownership (investment) quota of ith signatory
u_i = usage percentage of ith signatory
EC_i = effective cost of INTELSAT ownership and usage to i^{th} signatory

Development of the Model

As an investor in INTELSAT, a signatory is called upon to contribute his quota share of the capital required during a given period. That amount is given by q_iK. In return, he receives revenue distributions to reimburse him for his quota share of accounting depreciation, as well as to given him his share of return for use of capital. Thus, the revenue distribution to him will equal $q_i(D + R)$. Note that the revenue distribution concerns capital costs and return only and does not reimburse the investor for operating expenses, which are paid by users exclusively.

Furthermore, an investor has a net ownership or equity in INTELSAT (in absolute terms) which may increase or decrease during the period in question. His loss in equity is given by $q_i(D - K)$, that is, his quota share of the excess of depreciation over new capital contributions.

As a user, the signatory purchases a given number of communications channels at the prevailing price. The utilization charge, in turn, is set

---

The approach used in this appendix is originally due to William E. Snow, Jr., of the Communications Satellite Corporation, and a doctoral candidate at George Washington University. Any shortcomings in the particular development shown, however, are those of the present author.

at a level to recover total costs, that is, depreciation plus return to capital plus operating expenses:

$$P + \frac{D + R + O}{N} \tag{A. 1}$$

The signatory as user purchases u_iN units of utilization, for which he pays u_iNP dollars. But $u_iNP = u_i(D + R + O)$. (Equation A. 1 is only an accounting identity and not a cost function; it contains ex ante estimates ragarding the future levels of costs and traffic, on the basis of which a price is calculated.)

A signatory's effective cost as an INTELSAT owner-user, therefore, can be summed up as follows: it is equal to his loss of equity in INTELSAT, plus the cost of his capital contributions, less the revenue distribution he receives for depreciation and return on capital provided, plus whatever he pays for the capacity he purchases. The first three components of his effective cost are transactions as an owner, and the last represents his cost as a user. This is represented mathematically as follows:

$$EC_i = q_i(D - K) + q_i(K) - q_i(D + R) + u_i(D + R + O)$$

$$= R(u_i - q_i) + u_i(D + O) \tag{A. 2}$$

Equation A. 2 summarizes a signatory's financial interest in INTELSAT membership. The first term, $R(u_i - q_i)$, represents his interest as an investor in INTELSAT. If ownership is exactly equal to usage, that is, if $u_i = q_i$, the amount of R, that is, the level (and ultimately the rate) of return on capital, is irrelevant to him, since he gains as much as an investor as he pays out as user for whatever level of R. This conclusion must be modified if there is interfactor substitutability between the factor represented by D and the factor represented by O. In that case, even if $u_i = q_i$, the level of R is significant to the extent it fails to reflect domestic and international capital markets generally. If it does not, then INTELSAT will have failed to choose a factor combination which is cost-minimizing from the standpoint of each signatory's effective cost as shown in Equation A. 2. As noted, imbalances of ownership and usage were a problem during the interim INTELSAT agreements, but under the definitive arrangements they are periodically realigned. It is still possible, however, for nonmembers of INTELSAT to use the system under certain circumstances. If $u_i > q_i$, a member is a net user and has an interest in as low a rate of return for capital as possible. If $u_i < q_i$, the member is a net investor and has an interest in as high a rate of return as possible. The second term, $u_i(D + O)$, represents the member's interest as a user of INTELSAT. Thus any user, regardless of his level of ownership of the system, has an interest in keeping capital costs (as reflected by their depreciation schedule which is decided upon as a policy matter by the Board of Governors) as well as current operating expenses as low as possible. The present study determines, but does not consider for minimization, current operating costs as well as total costs. For a consideration of minimization of the present discounted value of investment costs in INTELSAT the reader is referred to another study by the present author.[1]

In conclusion, it should be stressed that this appendix has given a financial model for a signatory's membership in INTELSAT only. There is no reflection of a signatory's financial interests in wholesaling or retailing INTELSAT circuits to his own domestic market.

NOTE

1. Marcellus S. Snow, "Investment Cost Minimization for Communications Satellite Capacity: Refinement and Application of the Chenery-Manne-Srinivasan Model," Bell Journal of Economics and Management Science 6, no. 2 (Autumn 1975).

BIBLIOGRAPHY

Antitrust Bulletin 13 (Fall 1968). Symposium on regulation of communications.

Ashley, George E. "International Communications: What Shape to Come?" Journal of Law and Contemporary Problems 34, (Spring 1969), pp. 417-28.

Asimov, Isaac. "Toward the Global Village: The Fourth Revolution." Saturday Review, (October 24, 1970), pp. 17-20.

Averch, H., and Johnson, L. L. "Behavior of the Firm Under Regulatory Constraint." American Economic Review 52, (December 1962), pp. 1053-69.

Bain, Joe S. Industrial Organization. 2d ed. New York: John Wiley, 1968.

Bargellini, P. L., ed. Communications Satellite Systems. Progress in Astronautics and Aeronautics, no. 32.Cambridge, Mass.: M.I.T. Press, 1974.

_____. Communications Satellite Technology. Progress in Astronautics and Aeronautics, no. 33. Cambridge, Mass.: M.I.T. Press, 1974.

Baumol, William J. "Reasonable Rules for Rate Regulation," in Prices: Issues in Theory, Practice, and Public Policy. Edited by A. Phillips and O. E. Williamson. Philadelphia: University of Pennsylvania Press, 1967.

Baumol, William M., and Bradford, David F. "Optimal Departures from Marginal Cost Pricing." American Economic Review 60 (June 1970), pp. 265-83.

Baumol, William J., and Klevorick, Alvin K. "Input Choices and Rate-of-Return Regulation: An Overview of the Discussion." Bell Journal of Economics and Management Science 1, (Autumn 1970), pp. 162-90.

Belgium. Service du Premier Ministre. Secretariat d'Etat à la Politique et à la Programmation Scientifiques. L'Europe Spatiale. White Paper, April 1972.

Blevis, B. C., and Card, M. L. "The Implications of Satellite Technology for Television Broadcasting in Canada," in Communications Satellite Systems. Edited by P. Bargellini. Cambridge, Mass.: M.I.T. Press, 1974.

Boggiss, Ivor W. "Canada Domestic Satcom Orbits." Electronic News, November 13, 1973.

Bonbright, James C. "Fully Distributed Costs in Utility Rate Making," American Economic Review 51 (May 1961), pp. 305-12.

Borchardt, Kurt. Structure and Performance of the U.S. Communications Industry: Government Regulation and Company Planning. Boston: Harvard Business School, 1970.

Brazil. Conselho Nacional de Pesquisas. Comissão Nacional de Atividades Espaciais. Project SACI. Report no. 3. Technical Report LAFE-104. February 1970.

Brown, Murray. On the Theory and Measurement of Technological Change. Cambridge: Cambridge University Press, 1966.

Canada, Ministry of Industry. White Paper on a Domestic Satellite Communication System for Canada. March 28, 1968.

Capron, William M., ed. Technological Change in Regulated Industries. Studies in the Regulation of Economic Activity. Washington: Brookings Institution, 1971.

Capron, William M., et al. Technological Change and Welfare in the Regulated Industries. Brookings Reprint 219. Washington: Brookings Institution, 1971.

Chitnis, E. V., Karnik, K. S., and Rao, B. S. "Satellite Television—A Formidable Tool for National Development." Mimeographed (Indian Space Research Organization, 1971.)

Clemens, Eli W. "Discussion." American Economic Review 53 (May 1963), pp. 482-85.

_____. "The Marginal Revenue Curve under Price Discrimination." American Economic Review 38 (June 1958), pp. 388-90.

_____. "Price Discrimination in Decreasing Cost Industries." American Economic Review 3- (December 1941), pp. 794-802.

Colette, Rene C. "The European Communication Satellite Program," in Communications Satellite Systems. Edited by P. Bargellini. Cambridge, Mass.: M.I.T. Press, 1974.

Colino, Richard R. "INTELSAT: Doing Business in Outer Space." Columbia Journal of Transnational Law (1967), pp. 17-60.

_____. "International Satellite Communications and Developing Countries." Journal of Law and Economic Development 31 (Winter 1968), pp. 8-41.

_____. "Use of the INTELSAT Satellite System for Project Apollo: A Possible Precedent for Television Distribution Requirements." EBU Review, B, 118B (November 1969), pp. 24-33.

Communications Satellite Corporation. Annual Report, 1964-75.

_____. European Office. European Report. Geneva, November 1, 1968, and January 1, 1970. Issued October 1972 and October 1973 as Geneva Office Handbook.

_____. Information Office. News Digest. Issued approximately twice weekly. 1965-75.

_____. Pocket Guide to the Global Satellite System. December 1969, December 1, 1970, April 15, 1971, June 1, 1973, and March, 1974.

_____. Report to the President and Congress. Issued annually. 1964-75.

Dalfen, Charles M. "The Telesat Canada Domestic Communications Satellite System." Stanford Journal of International Studies 5 (June 1970), pp. 84-113.

D'Arcy, Jean. "Toward the Global Village: Challenge to Cooperation." Saturday Review, October 24, 1970, pp. 24-73.

Davidson, R. J. "The European Telecommunications Satellite Programme," ESRO/ELDO Bulletin, November 1970, pp. 11-18.

Dixit, Avinash K. "On the Optimum Structure of Commodity Taxes." American Economic Review 60 (June 1970), pp. 295-301.

Ende, Asher H. "International Telecommunications: Dynamics of Regulation of a Rapidly Expanding Service." Law and Contemporary Problems 34 (Spring 1969), pp. 389-416.

L'Espace et la Communication: Colloque International. Proceedings of an international colloquium at Paris, March 29-April 2, 1971. Paris: Editions Chiron, 1971.

"Europe and Space: Facts and Prospects." Articles translated from Syntheses, no. 275 (May 1969).

Fanti, Piero. "European Satellite Rates for Television Service." EBU Review, B, 118B (November 1969), pp. 34-40.

Farrell, M. J. "In Defence of Public-Utility Price Theory." Oxford Economic Papers, new ser., vol. 10 (1958). Reprinted in Tuvey, R., ed. Public Enterprise: Selected Readings. Baltimore: Penguin Books, 1968.

Faulhaber, Gerald R. "On Subsidization: Some Observations and Tentative Conclusions." Holmdel, N. J.: Bell Telephone Laboratories, 1972.

Feldman, Nathaniel E., and Kelley, Charles M., eds. Communications Satellites for the 70's: Systems. Progress in Astronautics and Aeronautics, no. 26. Cambridge, Mass.: M.I.T. Press, 1971.

_____. Communications Satellites for the 70's: Technology. Progress in Astronautics and Aeronautics, no. 25. Cambridge, Mass.: M.I.T. Press, 1971.

Foch, René. "L'Europe et la Technologie: Un Point de Vue Politique." Les Cahiers Atlantiques 2 1970). Paris: L'Institut Atlantique. Available in English as "Europe and Technology."

France. Centre National de le Recherche Scientifique. Groupe de Travail sur le Droit de l'Espace. Les Télécommunications par Satellite: Aspects Juridiques. Paris: Editors Cujas, 1968.

Fuss, Melvyn A., and McFadden, Daniel L. "Flexibility versus Efficiency in Ex Ante Plant Design." Discussion paper no. 190, Harvard Institute of Economic Research, Harvard University, June 1971.

Galloway, Jonathan F. The Politics and Technology of Satellite Communications. Lexington, Mass.: Lexington Books, 1972.

Heiden, Edward Joseph. "Some Economic Problems of International Space Telecommunications." Ph.D. dissertation, Washington University, 1967.

Henderson, James M., and Quandt, Richard E. Microeconomic Theory: A Mathematical Approach. Economics Handbook Series. 2d ed. New York: McGraw-Hill, 1971.

Hiergeist, F., and DeMontlivault, J. L. "The Earth Segment in the Symphonie Project," in Communications Satellite Systems. Edited by P. Bargellini. Cambridge, Mass.: M.I.T. Press, 1974.

Horley, Albert L. "An Approach to Planning Investment in Telecommunications for Development." Stanford Journal of International Studies 5 (June 1970), pp. 114-37.

Hotelling, Harold. "The General Welfare in Relation to Problems of Taxation and of Railway and Utility Rates." Econometrica 6 (1938), pp. 242-69.

Hughes Aircraft Company. Vectors 12 (Fall/Winter, 1970).

Hughes, William R. "Scale Frontiers in Electric Power," in Technological Change in Regulated Industries. Edited by William M. Capron. Washington: Brookings Institution, 1971.

"Das INTELSAT-Consortium and Seine Satelliten-Projekte," Astronomie und Raumfahrt, no. 6 (1968), pp. 179-84.

International Telecommunications Satellite Organization (INTELSAT). Documents of meetings of Interim Communications Satellite Committee (ICSC Board of Governors CBG), and Advisory Subcommittees on Financial, Technical, and Contracting Matters, 1965-75.

_____. INTELSAT Progress Report. Issued annually.

_____. INTELSAT Tariff Manual. Issued 1970 and updated periodically.

Intriligator, Michael D. Mathematical Optimization and Economic Theory. Series in Mathematical Economics. Englewood Cliffs, N.J.: Prentice-Hall, 1971.

Irwin, Manley R. The Telecommunications Industry: Integration vs. Competition. New York: Praeger, 1971.

Jacobs, Stanley L. "Domestic Satellites: How Will They Affect U.S. Communications?" The Wall Street Journal, January 31, 1973, p.1.

Johnsen, Katherine. "Four Domestic Satcom Plans Authorized." Aviation Week & Space Technology, September 17, 1973.

Johnson, Leland L. "Joint Cost and Price Discrimination: The Case of Communications Satellites," in Utility Regulation, New Directions in Theory and Policy. Edited by William G. Shepherd and Thomas G. Gies. New York: Random House, 1966.

_____. "Technological Advance and Market Structure in Domestic Telecommunications," American Economic Review 60 (May 1970), pp. 204-08.

Johnston, J. Statistical Cost Analysis. New York: McGraw-Hill, 1960.

Kahn, Alfred E. The Economics of Regulation: Principles and Institutions. 2 vols. Institutional Issues. New York: John Wiley, 1970 and 1971.

Kale, P. P., Nickelson, R. L., and Sarles, F. W., Jr. "A Design for INSAT," in Communications Satellite Systems. Edited by P. Bargellini. Cambridge, Mass.: M.I.T. Press, 1974.

Keeler, Theodore E. "Airline Regulation and Market Performance." Bell Journal of Economics and Management Science 3 (Autumn 1972), pp. 399-424.

_____. "Railroad Costs, Returns to Scale, and Excess Capacity." Review of Economics and Statistics 56 (May 1974), pp. 201-08.

_____. "Railroad Costs, Returns to Scale, and Excess Capacity: A Neoclassical Analysis." Working Papers In Economics, no. 35, Berkeley: University of California, 1973.

Kestenbaum, L. "Competition in Communications." Antitrust Bulletin 16 (Winter 1971), pp. 769-87.

Kildow, Judith T. INTELSAT: Policy-Maker's Dilemma. Lexington, Mass.: Lexington Books, 1973.

Knopow, J. J. "Next Generation Communications Satellites," in Communications Satellite Systems. Edited by P. Bargellini. Cambridge, Mass.: M.I.T. Press, 1974.

Lago, Armando M. "Demand Forecasting Models of International Telecommunications and Their Policy Implications." Journal of Industrical Economics 29 (November 1970), pp. 6-21.

Lave, Lester B., ed. Technological Change: Its Conception and Measurement. Series in Mathematical Analysis of Social Behavior. Englewood Cliffs, N. J.: Prentice-Hall, 1966.

Leive, David M. "Regulating the Use of the Radio Spectrum," Stanford Journal of International Studies 5 (June 1970), pp. 21-52.

Lerner, Abba P. "On Optimal Taxes With an Untaxable Sector," American Economic Review 60 (June 1970), pp. 284-94.

Lessing, Lawrence. "Cinderella in the Sky." Fortune, October 1967 pp. 131-208.

Levin, Harvey J. "Organization and Control of Communications Satellites." University of Pennsylvania Law Review 113 (January 1965), pp. 315-57.

______. "Spectrum Allocation Without Market." American Economic Review 60 (May 1970), pp. 209-18.

Lipsey, R. G., and Lancaster, Kelvin. "The General Theory of the Second Best." Review of Economic Studies 24 (1965), pp. 11-32.

Litzenberger, Robert H., and Rao, Cherukuri U. "Estimates of the Marginal Rate of Time Preference and Average Risk Aversion of Investors in Electric Utility Shares: 1960-66," Bell Journal of Economics and Management Science (Spring 1971), pp. 265-77.

MacAvoy, Paul W., and Noll, Roger G. "Relative Prices and Regulated Transactions of the Natural Gas Pipelines." Bell Journal Of Economics and Management Science 4 (Spring 1973), pp. 212-34.

MacLellan, D. C. "Anticipated Developments in Communications Satellite Technology," in Communications Satellite Systems. Edited by P. Bargellini. Cambridge, Mass.: M.I.T. Press, 1974.

Maddox, Brenda. Beyond Babel: New Directions in Communications. New York: Simon and Schuster, 1972.

______. "The Connections: A Survey of Communications." The Economist, 232, (August 9, 1969), pp. 1-40.

Marglin, Stephen A. "The Opportunity Cost of Public Investment," Quarterly Journal of Economics 77 (May 1963), pp. 274-89.

Marsten, Richard B., ed. Communication Satellite Systems Technology. New York: Academic Press, 1966.

Marzin, Pierre. "Space Communication: European Projects." EBU Review, pt. B, 118B (November 1969), pp. 19-23.

McDonald, John. "The Comsat Compromise Starts a Revolution." Fortune, October 1965, pp. 128-212.

McFadden, Daniel. "Criteria for Public Investment: Comment." Journal of Political Economy 80 (November-December 1972), pp. 1295-1313.

McWhinney, Edward, ed. The International Law of Communications. Dobbs Ferry, N. Y.: Oceana Publications, 1971.

Meek, Ronald L. "An Application of Marginal Cost Pricing: The 'Green Tariff' in Theory and Practice." Journal of Industrial Economics 11 (July 1963), pp. 217-36.

_____. "An Application of Marginal Cost Pricing: The 'Green Tariff' in Theory and Practice." Journal of Industrial Economics 12 (November 1963), pp. 45-63.

Merewitz, Leonard. "Costs and Returns to Scale in U. S. Post Offices." Journal of the American Statistical Association 56 (September 1971), pp. 504-09.

Meyer, J. R., and Kraft, G. "The Evaluation of Statistical Costing Techniques as Applied in the Transportation Industry." American Economic Review 51 (May 1961), pp. 313-34.

Mickelson, Sig. "Communications by Satellite." Foreign Affairs 68 (October 1969), pp. 67-79.

_____. "Toward the Global Village: The First Eight Years," Saturday Review, October 24, 1970, pp. 21-23.

Miller, Merton H., and Modigliani, Franco. "Some Estimates of the Cost of Capital to the Electric Utility Industry, 1954-57." American Economic Review 56 (June 1966), pp. 333-91.

Mishan, Edward J. "Criteria for Public Investment: Reply." Journal of Political Economy 80 (November-December 1972), pp. 1314-16.

_____. "Criteria for Public Investment: Some Simplifying Suggestions." Journal of Political Economy 75 (April 1967), pp. 139-46.

Moore, Frederick T. "Economies of Scale: Some Statistical Evidence," Quarterly Journal of Economics 73 (May 1959), pp. 232-45. Reprinted in Mansfield, Edwin (ed.), Microeconomics: Selected Readings (New York: W. W. Norton & Co., 1971), pp. 125-36.

Moore, Spencer. "The Canadian Broadcasting Corporation and a Domestic Communications Satellite." EBU Review, pt. B, 118B (November 1969), pp. 72-75.

Nelson, James R. "Practical Applications of Marginal Cost Pricing." American Economic Review 53 (May 1963), p. 480.

Peck, Merton J. "The Single-Entity Proposal for International Telecommunications." American Economic Review 60 (May 1970), pp. 199-203.

Peck, Stephen C. "Alternative Investment Models for Firms in the Electric Utility Industry." Bell Journal of Economics and Management Science 5 (Autumn 1974), pp. 422-58.

Pelton, Joseph Neal. Global Communications Satellite Policy: INTELSAT, Politics and Functionalism. Mt. Airy, Md.: Lomond Books, 1974.

______. "An Investigation of the International Telecommunications Satellite Consortium—The Prospects for Functionalist Development." Ph.D. dissertation, Georgetown University, 1972.

Planning for a Planet: An International Discussion on the Structure of Satellite Communications. Report of an international conference at Cap Ferrat, France. New York: Carnegie Endowment for International Peace and Twentieth Century Fund, 1971.

Posner, Richard A. "Natural Monopoly and Its Regulation." Stanford Law Review 21 (February 1969), pp. 548-643.

Riegel, O. W. "Communications by Satellite: The Political Barriers." Quarterly Review of Economics and Business 11 (Winter 1971), pp. 23-35.

Ropelewski, Robert R. "Europe L3S Launcher Start Set for 1974." Aviation Week & Space Technology, October 1, 1973, pp. 34-36.

Ruggles, Nancy. "Recent Developments in the Theory of Marginal Cost Pricing." Review of Economic Studies 17 (1949-50), pp. 107-27. Reprinted in Tuvey, R. (ed.), Public Enterprise: Selected Readings. (Baltimore: Penguin Books), 1968.

Sarabhai, Vikram. "Television for Development." Paper presented at Conference of Society for International Development, Delhi, November 14-17, 1969.

Schramm, Wilbur, et al. The New Media: Memo to Educational Planners. Paris: UNESCO, 1967.

Schwartz, Herman. "Comsat, the Carriers, and the Earth Stations: Some Problems with 'Melding Variegated Interests.'" Yale Law Journal 76 (January 1967), pp. 441-84.

_____. "Governmentally Appointed Directors in a Private Corporation—The Communications Satellite Act of 1962." Harvard Law Review 79 (December 1965), pp. 350-64.

Scrupski, Stephen E. "Marisat Partners Seek Accord." Electronics, June 13, 1974.

Shillinglaw, Thomas L. "The Soviet Union and International Satellite Telecommunications." Stanford Journal of International Studies 5 (June 1970), pp. 199-226.

Shute, John V. "A Domestic Satellite Communications System for the United States." EBU Review, pt. B, 118B (November 1969), pp. 67-71.

Silberman, Charles E. "The Little Bird That Casts a Big Shadow," Fortune, February 1967, pp. 108-226.

Simmons, N. "The United Kingdom Program of Communication Satellites," in Communications Satellite Systems. Edited by P. Bargellini. Cambridge, Mass.: M.I.T. Press, 1974.

Snow, Marcellus Scowcroft. "Cost Structure and Charging Policy in International Commercial Satellite Communications." Ph.D. dissertation, University of California, Berkeley, 1974.

_____. "Investment Cost Minimization for Communications Satellite Capacity: Refinement and Application of the Chenery-Manne-Srinivasan Model." Bell Journal of Economics and Management Science 6, no. 2 (Autumn 1975).

Solomon, Ezra. "Alternative Rate of Return Concepts and Their Implications for Utility Regulation," Bell Journal of Economics and Management Science 1 (Spring 1970), pp. 65-81.

Squire, Lyn. "Some Aspects of Optimal Pricing for Telecommunications," Bell Journal of Economics and Management Science 4 (Autumn 1973), pp. 515-25.

Stanley, Kenneth Bishop. "International Telecommunications Industry: Interdependence of Market Structure and Performance Under Regulation." Land Economics 49 (November 1973), pp. 391-403.

______. "Market Structure and Investment Behavior in the International Telecommunications Industry." Ph.D. disseratation, Michigan State Univeristy, 1971.

Stigler, George J. "The Theory of Economic Regulation." Bell Journal of Economics and Management Science 2 (Apring 1971), pp. 3-21.

Stockholm International Peace Research Institute. Communications Satellites, Stockholm Papers, no. 1, October 1968.

Switzerland. Federal Council. "Message du Conseil Fédéral Concernant l'Approbation de l'Accord Relatif à l'Organisation International de Télécommunications par Satellites 'INTELSAT,'" December 13, 1971, no. 11083.

Tassin, Jacques. L'Europe Spatiale. Paris: DeNoël, 1970.

Theil, Henri. Principles of Econometrics. New York: John Wiley, 1971.

Twentieth Century Fund Task Force on International Satellite Communications. Communicating by Satellite. Background paper by Paul L. Laskin. New York: Twentieth Century Fund, 1970.

______. The Future of Satellite Communications: Resource Management and the Needs of Nations. Second Report of the Twentieth Century Fund Task Force on International Satellite Communications. Appendixes by Walter R. Hinchman and D. A. Dunn. New York: Twentieth Century Fund, 1970.

United Nations. Economic and Social Council. Communication in the Space Age: The Use of Satellites by the Mass Media. Paris: UNESCO, 1968.

U.S. Congress. An Act to Provide for the Establishment, Ownership, and Regulation of a Commercial Communications Satellite System. Public Law 87-624, 87th Cong., 2d sess., August 31, 1962. [Communications Satellite Act]

U.S. Department of State. Treaties and Other International Acts Series 5646. International Telecommunications Satellite Consortium. Agreement Between the United States of America and Other Governments, with Special Agreement and Supplementary Agreement on Arbitration. Washington, August 20, 1964, and June 4, 1965.

______. Treaties and Other International Acts Series 7532. International Telecommunications Satellite Organization (INTELSAT). Agreement

Between the United States of America and Other Governments, and Operating Agreement, with Annex Concluded by Certain Governments, and Entities Designated by Governments. Washington, August 20, 1971.

U. S. Office of the President. Office of Telecommunications Policy. International Facilities Study. May 1971.

U. S. President's Task Force on Communications Policy. Final Report, December 7, 1968.

U. S. "Memorandum for the Honorable Dean Burch, Chairman of the Federal Communications Commission." Mimeographed. January 23, 1970.

Valters, Erik N. "Perspectives in the Emerging Law of Satellite Communications." Stanford Journal of International Studies 5 (June 1970), pp. 53-58.

Wald, H. P. "The Classical Indictment of Indirect Taxation." Quarterly Journal of Economics 59 (1945), pp. 577-96.

Walters, A. A. "Production and Cost Functions: An Econometric Survey." Econometrica 31 (January-April 1963), pp. 1-66.

Weil, Gordon L. Communicating by Satellite: An International Discussion. Report of an international conference held in Talloires, France. New York: Carnegie Endowment for International Peace and Twentieth Century Fund, 1969.

Williamson, Oliver E. "Economies as an Antitrust Defense: The Welfare Tradeoffs." American Economic Review 58 (March 1968), pp. 18-36.

Winston, Donald C. "European Space Agency Shift Stalled." Aviation Week & Space Technology, April 22, 1974.

Worcester, D. A., Jr. "Justifiable Price 'Discrimination' under Conditions of Natural Monopoly: A Diagrammatic Representation." American Economic Review 38 (June 1948), pp. 382-88.

Yaged, Bernard, Jr. "Economies of Scale, Networks, and Network Cost Elasticity." Holmdel, N. J.: Unpublished manuscript, Bell Telephone Laboratories, n. d.

Zajac, E. E. "A Geometric Treatment of Averch-Johnson's Behavior of the Firm Model." American Economic Review 60 (March 1970), pp. 117-25.

_____. "Some Preliminary Thoughts on Subsidization." Unpublished manuscript, Murray Hill, N. J.: Bell Telephone Laboratories, 1972.

INDEX

ABOUT THE AUTHOR

MARCELLUS S. SNOW is assistant professor of economics at the University of Hawaii. Before assuming that position in 1974, he completed a Ph.D. in economics at the University of California, Berkeley, during which time he taught microeconomics at California State University, San Francisco.

Dr. Snow was an assistant in the INTELSAT Affairs Division and subsequently in the Special Projects Division of the Communications Satellite Corporation, Washington, D. C., from 1969 to 1971; he served briefly in its Geneva, Switzerland office during 1973. His present areas of interest include econometrics and applied microeconomic theory. Among his publications is an article on communications satellites in the Bell Journal of Economics.

Prior to his Berkeley degree, Dr. Snow received an honors B. A. from the University of Utah; an M. S. from Massachusetts Institute of Technology; and an M.A. from the School of Advanced International Studies, The Johns Hopkins University.